The Battle of Columbus
April 16-17, 1865

The Battle of Columbus
April 16-17, 1865

by
J. David Dameron

Southeast Research Publishing, LLC

© 2017 by J. David Dameron

All rights reserved.

No part of this publication may be reproduced, stored in a retrieval system, or transmitted, in any form or by any means, electronic, mechanical, photocopying, recording, or otherwise, without the prior written permission of the publisher.

Printed in the United States of America.
Cataloging-in-Publication Data is available from the Library of Congress.

ISBN 13: 9780692884089

10 09 08 07 06 5 4 3 2

Picture Credit
Front Cover Image- Library of Congress

Unidentified soldier in first lieutenant's uniform, red sash, leather gauntlets, and spurs with cavalry sword.

Quarter-plate, hand-colored tintype. Liljenquist Family Collection, Prints and Photographs Division, Library of Congress [Digital ID # ppmsca-30998].

Published by
Southeast Research Publishing, LLC

This book is respectfully dedicated to the late Judge George Greene (1950-2014) of Phenix City, Alabama.

CONTENTS

Acknowledgements & Introduction   iv

Chapter 1 Battle Overview   1

Chapter 2 Wilson's Raid   11

Chapter 3 Columbus   56

Chapter 4 Confederate Bastion   78

Chapter 5 The Battle   129

Chapter 6 The Aftermath   176

Chapter End Notes   218

Appendix A: Captured Flags   230
Appendix B: Battle Analysis   232
Appendix C: Image Credits   251
Appendix D: Select Unit Histories   255

Bibliography   285

Index   299

Acknowledgements
and
Introduction

The author wishes to express his sincere gratitude to the people of Columbus, Georgia and Phenix City, Alabama. Southern hospitality is simply a way of life in the Chattahoochee Valley and the author was provided invaluable assistance from everyone asked and at every opportunity- Thank You!

On so many occasions, their hospitality and eagerness to assist in uncovering the details and pulling together facts made the research fun and a shared experience for all concerned.

While artifacts, anecdotes, and some technical details of the battle (from the Union viewpoint) are mentioned in several scholarly works, the battle receives little more than footnotes in the recorded history of the Civil War. The battle itself was simply overshadowed by larger events such as the rapid fall of the Confederacy, the death of President Lincoln, and due to slow communications, the realization that the war was truly over.

Additionally, after swiftly crushing resistance and destroying Columbus, the Union cavalry continued its mission sweeping on towards Macon, Georgia, leaving the region in a state of shock and devastation. All these events coupled with the vast destruction of the South, myriad uncertainties, economic ruin, sudden social upheavals, and the realities of a destroyed way of life, all combined to quickly diminish matters associated with the battle. Following the war, the former Confederate soldiers and the Southern people focused on survival and reconstruction.

Accordingly, many of the Southern records associated with the battle were tossed aside, lost or destroyed along with the Confederate Army. Nonetheless, some of the recorded facts were secured in local (public and private) collections and passed along through the generations. but lost to time as the South focused on Reconstruction. Pulling together these surviving accounts is a challenge as they reside in private letters, journals, and old documents locked away for many years.

As a case in point, the Confederate Order of Battle in Columbus on April 16, 1865, has heretofore, never been accurately reported. However, through surviving local records such as the Mayor's municipal and hospital records, detailed lists of the Confederate units posted and the names of casualties (both Union and Confederate) provide details that have long been overlooked.

These insights have never been reported in previous publications of the Civil War. The relevance of these records is key to understanding the battle, the combatants involved, and their disposition following the fight.

The Battle of Columbus was certainly unique in the annals of the Civil War and many of the details surrounding this event have remained a mystery for far too long.

Accordingly, the author wishes to express a special thank you to his friend and learned historian, the late Judge George Greene of Phenix City, Alabama. His private collection of original documents, letters, diaries, artifacts, detailed knowledge of the local history, and the Civil War era could fill volumes. Without his advice, assistance, and resources, the full details of the battle would have remained a mystery.

Chapter 1
Battle Overview

Often highlighted as being brutal and unnecessary, the Battle of Columbus was one of the last events in the long and violent American Civil War. The Union Cavalry Corps commanded by Brevet Major General James H. Wilson attacked the composite remnants of both Alabama and Georgia troops commanded by Major General Howell Cobb.

The industrial center of Columbus, Georgia was a target in a series of planned attacks in a campaign that had begun that spring. Sweeping eastward across Alabama and Georgia to eliminate Confederate resistance, destroy materiel and industrial facilities, "Wilson's Raid" was a brilliant Union success.

On April 16, 1865, the Union cavalry forces commanded by Brevet Major General James H. Wilson attacked the western earthwork defenses guarding the Confederate industrial center of Columbus, Georgia.

While the war had effectively ended with Lee's surrender of the Army of Northern Virginia on April 9, 1865, Wilson was attacking a region with severed lines of communications and he was uncertain of this rumored circumstance until days after the battle of Columbus.

Sweeping across Alabama from northwest to southeast in the spring of 1865, General Wilson's first major clash with the cavalry troops of General Nathan B. Forrest was at the Battle of Ebenezer Church on

April 1, 1865.

Wilson's Union cavalry then shattered resistance in Selma, Alabama on April 2nd, and intimidated the old Confederate capital of Montgomery into surrendering without a fight on April 12th. As the demoralized Confederates fled into Georgia, defenses were organized along the strategic bridges of the Chattahoochee River at Columbus, Georgia.

Columbus, Georgia was a valuable Confederate commodity as the town was a large industrial center. Except for the arsenals and manufacturing done at Richmond and Selma, Columbus was an invaluable Confederate lifeline providing pistols, swords, bayonets, shoes, uniforms, tents, buckets, and a multitude of accoutrements.

It also served as a Naval port and shipbuilding facility. Furthermore, Columbus served as the regional hub for cotton warehousing and transshipment via the Chattahoochee River, which empties southward into the Gulf of Mexico.

As survivors and refugees from Selma and Montgomery, Alabama fled towards Georgia, they carried word of the impending arrival of Union forces. Confederate cavalrymen of General Abraham Buford's command clashed with Wilson's corps repeatedly in attempts to delay the massive thrust of the Union forces.

These brief delays were little more than a nuisance for Wilson's cavalry, but nonetheless, time was critical for the defenders of Columbus as Negroes were forced to dig a "tete de pont" at the northernmost wagon bridge across the Chattahoochee River.

While there were two wagon bridges and one railroad bridge that could support conveyance of large forces, the main, 14th Street Bridge connecting Girard, Alabama (now Phenix City) with Columbus, Georgia was the focus of the Confederate defense.

For the Union marauders, the bridges at Columbus would have to be secured or General Wilson would be detoured at least twenty miles to the north or south of his target. The Confederate defenders were determined to keep the Union raiders out of Georgia.

In official reports recorded by the United States Army, the hostilities at Columbus have been expressed as a "battle," a "skirmish," and as an "action." Thus, the divisive debate concerning its actual prominence in history creates some controversy. With a full Cavalry Corps, Wilson's raiders pitted a force of nearly 13,500 men against an estimated Confederate force of 3,200 entrenched soldiers armed with 24 pieces of light artillery in fortified positions.

The military objective of the Union force was to destroy the industrial capabilities in Columbus, Georgia and continue to Macon. Meanwhile, the Confederate forces were determined to contain the Union marauders in Alabama (west of the Chattahoochee) and defend Columbus from their enemy.

Wilson's Union Cavalry Corps was well equipped, well trained, readily supplied, and overall an extremely efficient organization. He and his unit commanders were also extremely efficient and experienced. One of Wilson's commanders was Emory Upton, who later wrote the Army's basic manual of tactics and drill

(Upton's Tactics), which served the Army for several decades.

Major General Howell Cobb's Confederate units were an ad hoc organization that was neither well trained nor well organized. In fact, Cobb deferred overall command to the local defense force commander, Colonel Leon Von Zinken.

Von Zinken was an experienced and capable officer, but he was forced to defend a large industrial complex, housed within a city, and filled with civilian inhabitants. Furthermore, his units were remnants of their original organizations and the bulk of his men were reserve forces and either too young or too old for field service. Many of these men were members of the "Invalid Corps" and again, ill-prepared to fight.

While Von Zinken did have elements of Buford's cavalry available, they were used primarily to guard the flanks of the city to the north and south. Additionally, while the defenses contained fortified positions, the Confederates lacked a sufficient quantity of men to man them all.

To place the combatant forces in proper context, the Union forces were far superior in both numbers and capabilities than their Confederate antagonists. Yet, the Confederates did have forewarning of the impending arrival of their foe, and they were situated in a defensive position awaiting the imminent attack.

For Wilson, the battle was yet another step in his overall campaign goal to sweep across to Macon, Georgia destroying all enemy resistance in the region. Having attacked Selma, Alabama in a similar matter two weeks prior to this assault provided him with valuable insight as to how the Confederates would fight.

Again, after his victory at Selma, Wilson planned a similar assault on Montgomery, but the weakened Confederate forces and local civilians knew they could not withstand his forces, and they surrendered. Wilson may have hoped that Columbus would surrender as well, but he was just as prepared to fight if required.

In Columbus, the Confederates were determined to defend their resources and repel Wilson's assault, although they were all uncertain as to how they would perform with such a beleaguered and ill-prepared composite organization.

Both sides had adequate time to prepare for battle and there was no shortage of arms and ammunition. While the Confederates lacked sufficient numbers of men to fill all fortifications as desired, they consolidated their men in defense of the bridges. Accordingly, the main bridge was heavily defended and they were all prepared for immediate destruction (upon command) by soaking them with turpentine and cotton.

This measure was to prevent the use of the bridges by the Union forces, and further north at West Point, Georgia, General Tyler was entrenched and prepared to defend the crossing at that location. Wilson desired to capture the bridges at one locale or both, cross the Chattahoochee, and destroy the targets, eliminate all resistance, and continue his raid through Georgia.

Additionally, Von Zinken had organized his Confederate artillery into excellent positions, and apparently, the Confederates were confident in their ability to repel the attacking cavalry with long-range cannon-fire.

Several key personalities had a direct impact on both strategy and tactics in this battle. The Union employed an elite unit comprised of cavalrymen led by young generals that constantly pushed hard. Their aggressive nature coupled with innovation and initiative played key roles in their assault.

Likewise, but in the exact opposite context, the Confederate forces were led by leaders suffering from a lack of resources and demoralized forces. While both Generals Cobb and Buford were present in the defense of Columbus, Colonel Von Zinken was charged with organizing and leading the fight.

Buford's units were nearly decimated when they reached Columbus and they were exhausted having skirmished with Wilson for weeks in attempts to delay him. General Cobb had simply brought the remnants of units available from Georgia to assist in the defense of Columbus.

Viewing the objective of Columbus, Georgia as a target from the Union position in Montgomery, Alabama, Wilson had several options available. His forces could swing around to the north or south of Columbus, and attempt crossings at West Point, Georgia (north) or Eufaula, Alabama (south).

Either course would take several more days of travel than a direct route and there was a scarcity of habitation, food, forage, etc., in these rural regions to sustain his corps. Additionally, rapid and aggressive movement is the hallmark of cavalry operations, thus the direct route was the most desirable course of action.

Furthermore, the terrain (swampy lowlands and thick brush) channeled his movement along the main wagon road that led eastward through the small towns of Tuskegee and Crawford, Alabama. When Wilson reached Columbus, he would have to perform a reconnaissance and make a detailed plan of attack based on the situation as presented. His Union force was well prepared for this action, and a siege of Columbus was not an option.

From the Confederate perspective, the hastily organized force had to prepare fortified defenses and capitalize on their strength using artillery. The Union force was much too large to engage beyond Columbus except for hit and run delaying actions performed by the Confederate cavalry. The only real alternatives available involved specific placement of the limited manpower available to defend the city and its industries.

Confederate engineer, General Jeremy F. Gilmer supervised construction of 10 forts around Columbus in 1863, but most of these were too far (1-2 miles) beyond the river and the Confederates lacked adequate personnel to man them. A series of defense lines placed in depth (exterior, intermediate and interior) could have been constructed (like Richmond), but the Confederates chose not to do this, citing a lack of personnel.

The Confederates focused their defensive positions in several earthwork fortifications and connecting trench lines close to the river. All manned positions were within ½ mile west of Columbus, Georgia and oriented to defend the main wagon bridge (14th Street

Bridge) in a "tete de pont." These defensive positions were located on the Alabama side of the river and situated primarily to the northwest on Summerville Road, along the hills overlooking the river below.

Several established dirt roads provided easy movement to the bridges that crossed the Chattahoochee River at Columbus. In the north, the Summerville Road led downward to the main (14th Street Bridge) and in the south end of town, the Crawford Road led from Crawford, Alabama eastward to the Dillingham Street Bridge. These two bridges were about ¼ mile apart.

The Union cavalry approached Columbus from Crawford, Alabama, thus initially the lower (southern) Dillingham Street Bridge was in a direct path of assault. The lower bridge was lightly defended and prepped for destruction. Along the flanks (north and south) of the bridges, the Confederates placed cavalry to watch for attempted Union incursions outside of the defensive perimeter.

The Chattahoochee was too deep and too wide for a forced crossing; thus, Wilson would have to employ a pontoon bridge (he had adequate resources for this) or win a bridge to cross. There was another bridge (foot bridge) 3 miles north of Columbus at the Clapp's Factory. This bridge was prepared for immediate destruction also. The nearest wagon bridge beyond the three (2-wagon, 1 Railroad) at Columbus was 30 miles north at West Point, Georgia, and this location was also heavily defended.

Meteorological conditions on April 16-17 presented clear, comfortable springtime conditions for both sides (warm day, cool night). The most critical element that impacted both sides was the lunar conditions, which presented negligible luminosity. The level of darkness provided excellent conditions for a "fog of war," which either side was free to exploit.

Columbus, Georgia lies in the heart of the Chattahoochee Valley where the Appalachian Mountain chain ends its southernmost point. This natural fall line geographically presents rolling hills to the northwest, and predominantly flat lands to the southwest.

The maximum elevations in this area reach 450 feet. As with most cities in 1865, much of the timber in the immediate area was removed for lumber and fuel, however, the woods along the riverbank, above and below the city were filled with pines and several varieties of small hardwood trees.

General Wilson's Union Cavalry Corps overwhelmed the entrenched Confederate forces at Columbus. The engagement represented the final major battle of the war.

While there were several clashes in remote locations following the Battle of Columbus, they were of no tactical consequence, as the Confederacy no longer existed.

Nonetheless, the final major battle of the war and the events leading to that engagement present an interesting yet often overlooked chapter in the American Civil War.

CHAPTER 2
WILSON'S RAID

One of the greatest strengths of the Confederate Army was its cavalry. Led by men such as J.E.B. Stuart, John Hunt Morgan, and Nathan Bedford Forrest, these seemingly fearless raiders rode hard and fast, delivering sudden, deadly blows against their Union adversaries. The Confederate use of cavalry as a mobile offensive strike force was very different from the pre-war military thought that cavalry should be limited to reconnaissance, screening missions, and escort service.

In 1863, after learning costly lessons in battle, the U.S. Army embraced the cavalry as an autonomous combat arm, and an independent fighting force. Transforming their tactics and doctrine to capitalize on the inherent strength of the cavalry was not only time consuming; it was also conducted while fighting a war. In the eastern theater of war, General Phillip Sheridan led the Union cavalry reorganization to emulate their Confederate counterparts who employed cavalry in offensive tactics.

While their mission was difficult, Sheridan and his cavalrymen quickly adapted to their new role as shock troops. One of Sheridan's finest horsemen was a young and dashing division commander named James Harrison Wilson. While he was only 27 years old, Wilson was a West Point graduate, a skillful leader, and a supreme cavalryman. He had also served as the Chief of the Army's Cavalry Bureau, and during the Vicksburg Campaign, Wilson served under General Ulysses S. Grant.

Thus, General Grant was well acquainted with Wilson's record of performance and capabilities, and based upon these personal observations, Grant selected Wilson for a special mission. His task was to lead a cavalry raid into enemy occupied Alabama and Georgia.

Wilson's mission would serve as the third Union cavalry raid into Alabama. The first mission had met with disaster.

From April 19 to May 3, 1863, Union Colonel Abel D. Streight struck northern Alabama with 1,700 men to sever the railroad lines supplying the Confederate Army of Tennessee. Within three weeks, the cavalry forces of Nathan Bedford Forrest, with only 500 men, ran them down, and captured or killed Streight and his entire raiding force.

The second raid followed the loss of Tennessee to the Union in 1864. This time, both Alabama and Georgia, were attacked as General Sherman's army moved southward in Georgia, a major Union cavalry raid struck at central Alabama. Again, this raid was focused on disrupting Confederate supply trains delivering war materiel from the industrial centers of Selma, Alabama and Columbus, Georgia.

Led by Major General Lovell Rousseau, Commander of the districts of Nashville and Tennessee, his cavalry troops attacked critical points along the Montgomery and West Point Railroad. Moving deeper into the south than Colonel Streight's raid, Rousseau cut a swath from northwest to southeast across Alabama, from Decatur to Opelika, with 2,500 cavalrymen.

The raid was swiftly executed between July 10 and 22, 1864. It was one of the most brazen and successful cavalry raids of the war and it caught the Confederacy by surprise.

The raid achieved great success in its mission to disrupt and destroy. The Union raiders destructed thirty miles of critical Confederate railroad tracks, which emboldened the Union command to carry out even larger operations into the deep south.

By severing the supply trains feeding the Confederate defenses of Atlanta, Rousseau's raid directly supported Sherman's successful victory at Atlanta, in September of 1864.

The Union raid also helped to consolidate its forces in key points such as Huntsville, Alabama and the Union western army was effectively increased by 100,000 men. For the Confederacy, the raids in Alabama and Georgia were highly disruptive. Portions of Alabama and Georgia were now occupied by Union invaders. Refugees and Confederate wounded moved southward and resources were rapidly being depleted.

Capitalizing upon the successful summer cavalry raid led by General Rousseau, an even larger cavalry incursion into the deep south was planned by Generals Grant and Sherman. In October of 1864, General Grant promoted Brevet Major General James H. Harrison and transferred him to command all Union cavalry forces in Mississippi.

General Wilson reported initially to General William T. Sherman who had taken Atlanta and would soon embark on his infamous "March to the Sea." According to Wilson, "General Sherman, after issuing all the necessary instructions and unfolding his plans for the operations of the new cavalry corps, generously added:

> *Do the best you can with it, and if you make any reputation out of it I shall not undertake to divide it with you. Thus, the paper organization had its origin; but in as much as most of the force was dismounted and detachments of it were scattered from east Tennessee to southwestern Missouri, much the greater part of the real work of reorganization had yet to be done.* (B&L, vol. 4, p. 465)

Brevet Major General James H. Wilson

    Wilson's command was designated as the "Cavalry Corps of the Military Division of the Mississippi," and the young officer would command many men much older than himself. He also ascended to a position that made other commanders envious of his position. Wilson's assignment to the west was not only militarily challenging, it was politically sensitive as well.

    By placing the energetic young officer in the Army of the Mississippi, Grant planned to "put spurs" into the slow movements of the Union force in Tennessee, led by Major General George W. Thomas.

While General Thomas had won laurels as the heroic "Rock of Chickamauga," Grant considered Thomas to be "sluggish." Despite Grant's less than idealistic opinion of Thomas as a commander, he was a veteran of the Dragoons, an advocate of cavalry, and he eagerly supported Wilson's transformation of the cavalry.

Wilson quickly assessed the situation and went to work. Within a few short weeks, despite the constant threat of the enemy, Wilson had re-organized three full divisions and prepared them for battle. Kilpatrick's division of cavalry was detached and accompanied General Sherman on his campaign, but ultimately Wilson's corps would swell to six full divisions numbering 34,000 officers and men.

In short order, Wilson gained the confidence of his new commander, his peers, and most importantly, his men. While fighting in Tennessee, Wilson also gained the respect and disdain of his enemies.

At the battles of Franklin and Nashville, Wilson's Cavalry Corps achieved great success fighting against the Army of General John B. Hood and the Confederate "Wizard of the Saddle," General Nathan Bedford Forrest.

Following the Battles of Nashville and Franklin, the Confederate Army of Tennessee was in tatters. Remnants of Confederate units led by Generals John B. Hood retreated from Tennessee and escaped southward into Mississippi.

General Forrest also retreated into Mississippi where he reconsolidated his cavalry troops, and launched mounted patrols into Alabama to try to keep his enemy restricted to Tennessee.

The Confederate cavalry provided a defensive buffer for central Alabama and the western border region of Georgia, which remained unscathed by war. Here, the Confederacy thrived and supplied the Confederate war machine with badly needed stores of food, arms, and ammunition. This region was literally the "Heart of Dixie" as its major cities pulsed with industrial might. Selma, Alabama was a valuable southern city and well known for its manufacturing capabilities.

Montgomery, Alabama, the former capital of the Confederacy, was located just to the east of Selma. In addition to its symbolic importance, Montgomery was an important agricultural and materiel distribution center.

To the east of Montgomery, the Chattahoochee River flows southward through Girard, Alabama (present day Phenix City) and Columbus, Georgia. The Chattahoochee River continues to widen and deepen as it empties further south into the Gulf of Mexico at Apalachicola, Florida.

In the Confederacy, this manufacturing center was second only to Richmond in its industrial capacity. It served the Confederacy as a primary quartermaster supply center, foundries that forged cannons, and factories that made pistols, swords, bayonets and ammunition as well as shoes and uniforms.

Virtually everything needed to supply an army and navy was made there. In addition to the many items manufactured there, Columbus also housed a naval port and constructed gunboats. Additionally, these cities served as warehousing facilities for cotton, which the Confederacy used as a medium of exchange with European traders.

This resourceful region of the Chattahoochee Valley served as the richest and largest Confederate area that had not been directly threatened by Union assaults. The Union leadership knew the region warranted their attention, and in early 1865, they focused their sights accordingly.

The Cavalry Corps led by General James H. Wilson was selected as a mounted invasion force and readied for their mission. Re-supplies of horses, horseshoes, food, and ammunition were transferred from various units to strengthen Wilson's corps. Set to launch in the spring of 1865, Wilson prepared his corps and waited for the harsh winter weather to subside.

A dashing young Cavalry Officer, ready for battle with sash, spurs and sword.

While in winter quarters during January and February of 1865, Wilson focused on his unit's reorganization efforts, drilling his new corps, and preparing them for their next campaign. In writing the official reports of the battles of Franklin and Nashville, Wilson highlighted the importance of equipping his Cavalry Corps with the best materiel available. On December 21, 1864, General Wilson recorded:

If the operations just described have been of any avail in the recent campaign, it is due entirely to the concentration of the cavalry and its reorganization as a separate corps. I have, therefore, to request that the credit awarded it may be used to secure from the War Department the recognition of its separate existence as a corps, and an official approval of the measures already inaugurated for its efficiency.

With an opportunity to complete its organization, a full supply of Spencer carbines for the entire command, and we can take the field next spring with a force of cavalry fully competent to perform any work that may be assigned it. (OR 45-1 report # 194)

Wilson also wrote to General Alexander B. Dyer, the Union Chief of Ordnance, stating:

There is no doubt that the Spencer carbine is the best firearm yet put into the hands of the soldier, both for economy of ammunition and maximum effect, physical and moral. Our best officers estimate one man armed with it equivalent to three with any other arm. I have never seen anything else like the confidence inspired by it in the regiments or brigades, which have it.

A common belief amongst them is if their flanks are covered they can go anywhere. I have seen a large number of dismounted charges made with them against cavalry, infantry, and breast-works, and never knew one to fail. (January 2, 1865. Brig. Gen. A. B. Dyer, Chief of Ordnance (OR 45-2)

The .50 caliber Spencer carbine was a lever-action; breech-loading repeater that measured 39 inches in length and weighed just less than 9 pounds. The weapon held a tubular seven-round magazine filled with pre-made copper rim fire cartridges.

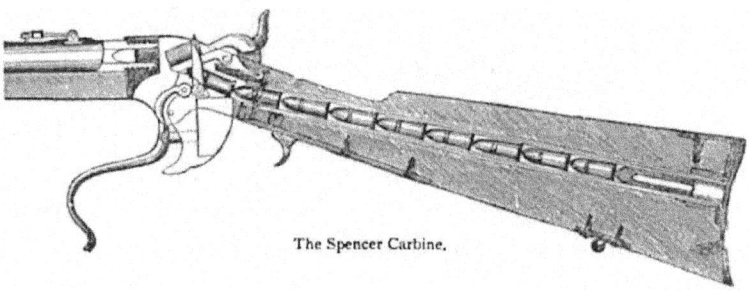

The Spencer Carbine.

The magazine was seated in the butt stock and a coil spring behind the rounds fed the cartridges into the breech. A simple operating lever dropped the breech block, extracted the spent round, and then by closing the lever, the next round was automatically fed, seated into the breechblock, and the weapon was ready to fire again. Soldiers simply cocked, aimed, and fired the weapon, and they could easily and accurately fire 7 rounds within 20 seconds.

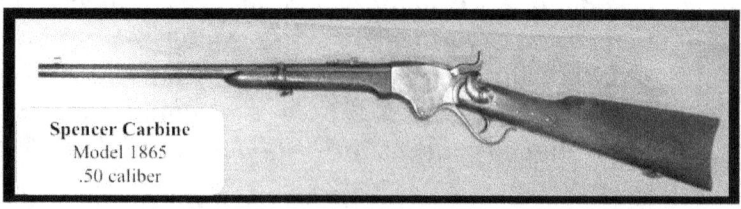

Spencer Carbine
Model 1865
.50 caliber

The Blakeslee cartridge box that General Wilson obtained for his troopers provided a dozen pre-loaded magazine tubes that greatly increased their battlefield lethality. While the range of the Union's primary weapon was less than their enemy, the Confederates were limited to just 3 rounds per minute, and they were encumbered by the mechanics inherent with muzzle-loaders.

Two US Cavalrymen proudly display their "tools of the trade": Spencer carbines, cavalry sabers and six-shot revolvers.

In addition to their Spencer carbines, each trooper carried a cavalry saber and a six-shot revolver. Again, while the range was short, the pistol provided yet another rapid-fire weapon for close combat, and most soldiers could accurately fire 6 rounds within 10 seconds.

In March of 1865, the grand Union strategy orchestrated by General U.S. Grant was set into motion. As the frigid weather loosened its icy grip on the southland, Wilson's Cavalry Corps assembled in the northwest corner of Alabama, at Gravelly Springs.

Southward across the Tennessee River and into the "Heart of Dixie," Wilson's Cavalry Corps eagerly sought to prove the lethality of their unique organization in Alabama.

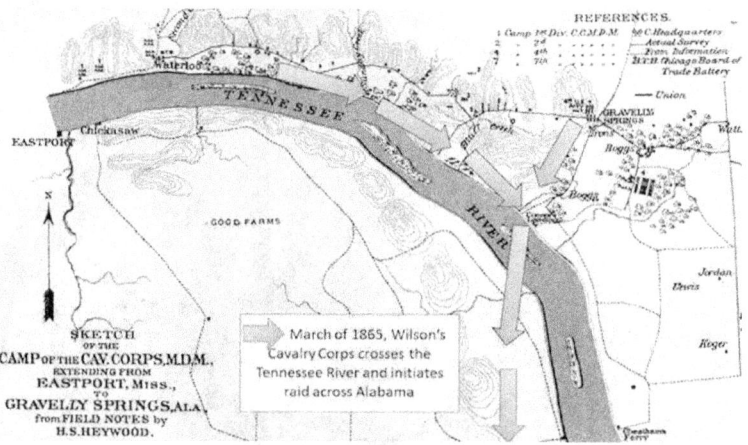

Wilson's camp and river crossing, March of 1865

By combining the swift mobility inherent in cavalry with the lethality of the Spencer carbine, Wilson's Corps would enjoy a combat overmatch previously unattainable in the annals of military history.

Armed with a cavalry saber, a six-shot revolver, and a Spencer seven-shot repeating carbine, the troopers of Wilson's Corps were afforded a force multiplier of seven to one.

These strengths were also enhanced by the fact that Wilson's men were well-trained, battle-hardened veterans, and their morale was high.

In stark contrast to their Union foe, morale in the Confederacy was reaching its ebb tide. Throughout the Confederacy, the grim reality of a long war took its toll and their beleaguered forces were deserting at an alarming rate. Especially demoralized were the men of Hood's badly beaten Army of Tennessee.

Even their commander who had given his all to the Confederacy lacked the motivation to continue the fight. John Bell Hood had lost an arm and a leg to the cause he held so dear, and after the Tennessee campaign he relinquished command of his shattered army.

Lieutenant General Richard Taylor, a highly-respected officer and the son of former president, Zachary Taylor, succeeded Hood as the commander of a newly formed department that included Mississippi, Alabama, and east Louisiana.

While his authority also extended into the Military Department of Georgia, this was a very politically sensitive issue as Governor Joe Brown fought the Confederate authorities bitterly concerning military affairs in Georgia.

CSA Lieutenant General Richard Taylor

In fact, Governor Brown insisted that Georgia troops would remain stationed within the state borders to defend it from Union aggression. Governor Brown and Major General Howell Cobb maintained "control" of their militia, yet Cobb eagerly responded to CS authorities including General Richard Taylor. Once spring arrived, Taylor planned to concentrate his defenses elsewhere in reaction to the movements of the enemy.

Taylor also dispatched several units to Augusta, Georgia where they joined General Joseph Johnston's newly formed Army of Tennessee. Generals Johnston and Hardee forged a combined force that the Confederacy hoped could delay Sherman's army marching northward into Virginia. With Grant pressing Richmond from the northern and eastern flanks, and Sherman approaching from the South, Lee's Army of Northern Virginia would not be able to defend the Confederate capital for long.

While Taylor was well respected, and an adept leader, he knew the Confederacy was doomed. Reflecting upon the reality of the situation, Taylor recorded that the "last act of the bloody drama" was upon them, but the "duty of a soldier in the field is to continue to fight until stopped by the civil arm of his government." (Taylor, 218)

Taylor commanded a mere shell of a once mighty army but he was not a defeatist. Taylor was a professional soldier, and a proud southerner that commanded men who were hell-bent on fighting for "the cause." Yet, except for Forrest's cavalry, his army was limited to defensive combat as his command was decimated.

The only viable threat to Wilson's Corps in Alabama was the cavalry force commanded by Lieutenant General Nathan Bedford Forrest. Known as the Confederate "Wizard of the Saddle," Forrest was a skilled adversary, but his command had suffered great losses during the Tennessee campaign.

CSA Lieutenant General Nathan Bedford Forrest

In Tennessee, Wilson's Cavalry Corps had quickly matured into a lethal battle-hardened force, and it was the largest cavalry command ever assembled. While General Forrest and his gallant Confederate cavalrymen were proud and efficient warriors, they were awestruck and somewhat envious of Wilson's corps.

One of Forrest's Division commanders, Brigadier General James "J.R." Chalmers informed his commander that, "To 'learn wisdom from your enemy' is one of the wisest maxims of history. At Nashville, our enemy had a large force of cavalry, but instead of wasting its strength in the front, he kept it quietly in the rear of his infantry, resting and recruiting, until the time for action came, and then he moved it out fresh and vigorous with telling effect."

CSA Brigadier General James Ronald Chalmers

Chalmers then urged Forrest to model their own units like their Union counterparts, stating, "If we had time to reorganize, recruit, and fit up the command in a place where forage could be procured, we can whip the enemy's cavalry, and every man in your command is anxious that you should have a fair trial of strength with Major General Wilson." (OR 45-2, January 3, 1865)

In addition to the division commanded by Brigadier General Chalmers, General Forrest's reorganized Cavalry Corps of the Department of Alabama, Mississippi, and East Louisiana included two other partial divisions commanded by Brigadier Generals William H. Jackson and Abraham Buford. He also had remnants of two partial brigades led by Brigadier General Philip D. Roddey and Colonel Edward Crossland. The State of Alabama also provided Forrest with several hundred mounted local militia troops.

CSA Brigadier General William H. Jackson

General Wilson was eager to meet his adversary as well, and he focused on him intently. Previously, in late January one of Wilson's cavalry officers, Captain Lewis Hosea was sent as an emissary on a prisoner exchange mission to meet with General Forrest and "keep his eyes open."

When Captain Hosea met with General Forrest, he relayed General Wilson's compliments and his commander's wish that they may meet "upon some future occasion." Forrest responded to this comment in his slow southern dialect, stating:

Just tell General Wilson that I know the nicest little place down here, in the world, and whenever he is ready, I will fight him with any number from one to ten thousand cavalry and abide the issue. General Wilson may pick his men and I'll pick mine. He may take his sabers and I'll take my six shooters. (Wills, quoted in 301-02; Wilson, p. 184)

Forrest actively questioned Hosea about General Wilson and while he acknowledged that Wilson was better educated than he, and a professional military tactician, Forrest was eager to fight him. Captain Hosea's observations provided General Wilson with a great deal of information concerning his enemy and like him, Wilson was also eager to "abide the issue."

To assist General Forrest in the impending fray, General Taylor supported him as well as he could. In February, Forrest was promoted to the rank of Lieutenant General and authorized to bolster his strength by adding several more units to his command.

In addition to his Mississippi, Tennessee, and Texas cavalrymen, the units of Brigadier General Daniel W. Adams from the Military Districts of North and Central Alabama were added to Forrest's command.

Adams' command was like Major General Howell Cobb's local forces in Georgia as he commanded all the militia, conscripts, and virtually anyone that Alabama Governor Thomas H. Watts could provide him.

CSA Brigadier General Abraham Buford

Thus, by mid-March of 1865, the Confederacy had its own reorganized corps of cavalry. While Forrest's Cavalry Corps was comprised of stalwart, experienced veterans, they lacked sufficient time and resources to adequately prepare it. In a message to one of division commanders, Forrest expressed, "Spare no time, hasten to reorganize and fit up your command. We have no time to lose." (OR 49-1, p.994)

General Forrest's Staff Officers

On March 16, 1865, under the cover of darkness, Wilson's advanced scouts crossed the Tennessee River and began a reconnaissance into enemy territory. By March 22, the entire column was underway and the massive Union cavalry corps was invading Alabama. (Harrington Diary)

The Union Cavalry Corps led by General Wilson was comprised of 13,500 men organized into three divisions commanded by Brigadier Generals Edward M. McCook, Eli Long, and Emory Upton.

Brigadier Generals Eli Long and Edward M. McCook

General Wilson's Cavalry Corps had been tasked to deliver a "swift saber-thrust at the heart of the Confederacy," and it was plunging in, hard and fast. (Gilpin, 618)

Ultimately, their campaign known today as "Wilson's Raid" would wreak a wide swath of destruction across Alabama and into Georgia. Along their route lay the cities of Selma, Montgomery, and Columbus, whose citizens were busily preparing for their defense. Fortified by extensive earthworks, armed with artillery, and manned with as many men as they could muster, the city/ fortresses awaited their foe.

As citadels of the Confederate cause, these proud southerners continued their industrial mission, and tried not to reflect on what Sherman had done to Atlanta. After all, the Confederate cavalry had generally held the advantage over Union cavalry during the war.

General Forrest's famous maxim, "get there first with the most men," acknowledges the simple military truth that strength and speed are critical to success in battle, but the value of timely information is equally important.

On March 20, Major General Frederick Steele stationed in Pensacola, Florida moved his 12,000-man force northward on an expedition that would ultimately join with General Edward Canby (nearly 32,000 men) in an assault on Mobile, Alabama. This early-season movement was quickly detected by the Confederacy and it signaled a threat to Alabama in the south.

Forrest immediately dispatched Generals Roddey and Buford southward towards Greenville, Alabama to protect that area of Alabama from attack. Unbeknownst to Forrest, the Union incursion from the south was focused on Mobile in southwestern Alabama, but the Union movement from the north threatened Montgomery. Forrest also acknowledged the latter threat with a counter-action.

General Forrest's decision to split his command would prove to be a serious tactical error; however, he was doing his best with limited resources against overwhelming forces.

General James H. Clanton's brigade was the first unit to confront the approaching Union column at Bluff Springs near Pollard, Alabama. The resulting battle on the 25th of March was a rapid Confederate disaster as the Confederates brigade suffered several hundred casualties, which included the son of Alabama Governor, Thomas Watts, and Clanton himself. The remnants of Clanton's brigade returned northward to Selma where they rejoined Buford's division. (49-1. p. 310)

CSA Brigadier General James H. Clanton

Dispatching large numbers of Confederate cavalry southward proved to be costly to the defense of the region. In fact, this Confederate reaction to the initial threat played directly into the hands of General Wilson who moved his corps rapidly from the northwest deeply into Alabama, and exactly as planned.

The Union plan worked perfectly and even when General Taylor learned that Union columns were approaching simultaneously from different directions, he failed to grasp how large these forces were. On March 27, in a situation report to General Robert E. Lee, General Taylor stated, "My intention is to meet and whip these detached columns before they can advance far into the country or unite with each other." (49-2, p. 1160)

By the time the Confederates realized the northern column was Wilson's entire Cavalry Corps (nearly 13,500 men), Forrest's Cavalry Corps (approximately 7,000 men) was widely dispersed, and facing an ominous threat. Forrest reacted quickly and sent Generals Roddey and Buford back to the north in hopes of delaying the massive Union column invading that sector of the Alabama.

General Grant had launched a series of simultaneous assaults against the Confederate states, as he reinitiated his war of attrition. Alabama was a primary target, and the total invasion force threatening that state quickly grew to approximately 60,000 men and 20 ships of the U.S. Navy.

On March 30, at Elyton (present-day Birmingham) Wilson detached a brigade led by General John T. Croxton to attack Tuscaloosa in western Alabama, while the main force continued towards Montevallo. Wilson's lead division was commanded by a bold, but very religious young man named Emory Upton.

Brevet Brigadier General Emory Upton

Upton was just 25 years old, but had quickly risen to the rank of Brevet Brigadier General due to his incessant displays of soldierly courage coupled with a devotion to God and country that rivaled any religious patriot. Upton was a driven man, and his orders were to take Montevallo, which he did on the evening of March 31st.

In short order, Upton's men captured the town, destroyed Confederate property, attacked Roddey's cavalrymen, and took 100 of his best men as prisoners. Wasting no time, Upton continued driving on into Randolph, Alabama and halted there on the 1st of April.

Meanwhile, Forrest made a grave mistake by further dividing his forces. Forrest decided that if he could send one of his units around to the rear of Wilson's giant column, he could simultaneously attack the Union column from the front and rear before they could threaten Selma. Roddey's men continued to harass the invading Union column as it marched southward towards Selma.

CSA Brigadier General Philip D. Roddey

Yet again, Upton's cavalrymen thoroughly routed the Confederates cavalry at Randolph, and most importantly, they captured a courier with Forrest's detailed plans. Wilson learned from the courier's dispatches that Forrest had sent General Jackson's 3000-man division to swing around to the west and come up on Wilson's rear.

*1st Sergeant John C. Gammill,
3rd Regiment, 1st Brigade, Upton's Cavalry Division*

Reacting to this threat, Wilson dispatched another brigade led by General Edward McCook westward who could work with Croxton and contain Jackson in that region. Meanwhile, Wilson knew that Forrest's Cavalry Corps was widely dispersed and confused by the Union onslaught. Moreover, Wilson also realized that Forrest was just down the road from his massive corps, and in a vulnerable predicament.

General Wilson ordered his division "to allow him no rest, but push him toward Selma with the utmost spirit and rapidity.

These officers, comprehending the situation, pressed forward with admirable zeal and activity." (49-1, p.358-9)

General Forrest was determined to halt Wilson's assaulting column, and to do so, he established a line of defense at Ebenezer Baptist Church, six miles north of Plantersville, Alabama. Extending his troops along the north bank of Bogler's Creek, Forrest prepared to fight General Wilson. Forrest anchored his right on Mulberry Creek, and his left terminated along a steep, thickly wooded ridge.

Upon these heights, he placed four cannons to cover any approach via the Randolph road, and he emplaced two more cannons to cover the Maplesville road approach as well. Forrest had about 3,000 men, which included elements of Roddey's division, Armstrong's brigade, of Chalmers' division, Crossland's Kentucky brigade of Buford's division, and a 300-man infantry battalion led by General Daniel Adams. The Confederates also hastily placed obstacles in front of their line consisting of rail barricades and a slashing of pine trees.

On the afternoon of April 1, 1865, Generals Wilson and Forrest engaged in the combat they both so eagerly desired. Throughout the day, the Confederates had continued their harassment of the Union column, leapfrogging units until they fell back to Ebenezer Church. Here, General Long's division was in the lead as the Union column smashed into Forrest's line of defense.

While Forrest had established a defense as best as he could, it proved to be no match for the heavy column that attacked it. General Long assaulted Forrest with a mixture of dismounted forces supported by a mounted battalion that broke through the line on the Confederate right.

Long then committed four more companies of the 17th Indiana Cavalry into the fray. The charge was led by Colonel

Frank Long. General Wilson described Long as a "berserker of the Norseman breed, broad-shouldered, deep-chested, long-limbed... and "bearded like a pard (old English meaning leopard)." Nonetheless, the Confederate cannon fire was so thick, they were forced to retreat.

At the head of the charge, unable or unwilling to hear the Union bugler sounding retreat, Captain James D. Taylor and sixteen other troopers of the Seventeenth Indiana slashed through the line attacking directly into General Forrest and his personal escort. Their assault was so furious that General Forrest was forced to engage in personal combat.

Captain Taylor sliced General Forrest's left arm with his saber and repeatedly hacked at his head and shoulders until Forrest raised his pistol and killed him. All of Taylor's men were killed, wounded or captured, and Forrest was shocked at the ferocity with which these Union cavalrymen sought to kill him. Forrest later remarked that the zealous young Union cavalryman nearly killed him. The wound was Forrest's fourth in battle yet he continued the fight.

While General Long was crushing the Confederate right, General Alexander of Upton's division pressed forward on the enemy's left. With the massed Union corps pressing in depth along the entire length of the line, in less than one hour, Forrest's men were overwhelmed and forced to retreat toward Selma.

Brigadier General Andrew J. Alexander, Commander, 2nd Brigade of Upton's 4th Division

General Winslow's brigade pursued the retreating Confederates, but the Union assault halted as the sun was setting, and the Union column reached within nineteen miles of Selma.

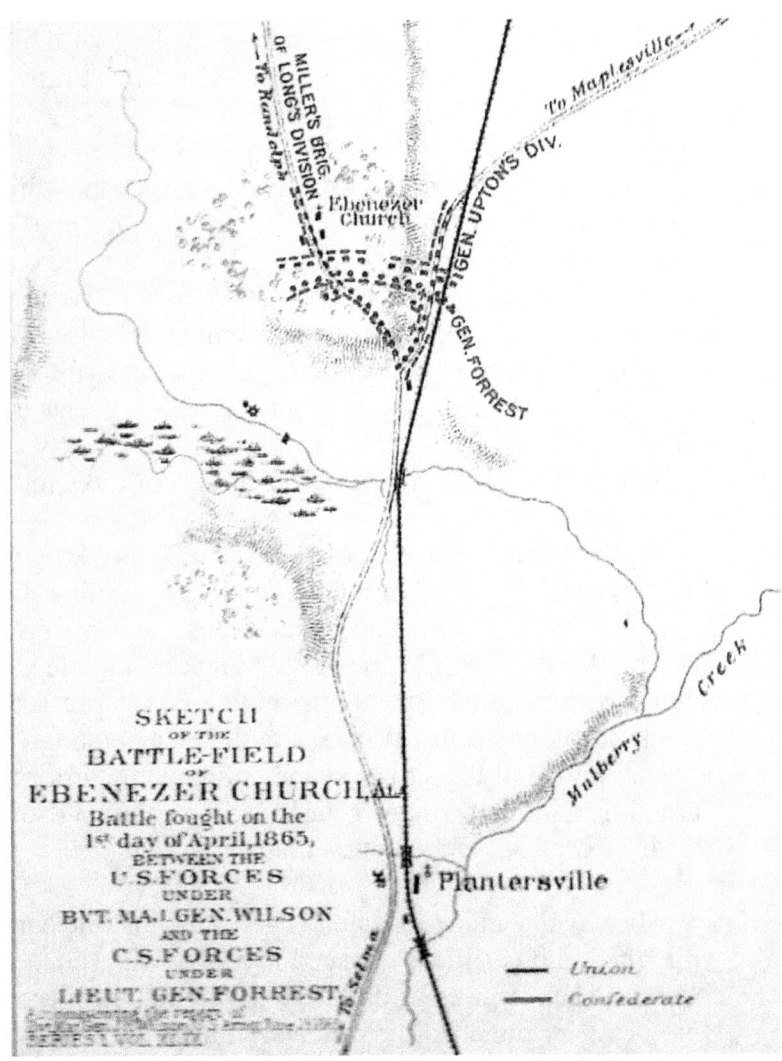
Battlefield Map- Ebenezer Church

At the Battle of Ebenezer Church, Wilson lost 52 casualties, but Forrest lost 3 cannons, three hundred men, and he nearly lost his own life. General Forrest spent the night regrouping his units and in the morning of April 2, he raced into Selma. Upon his arrival, Forrest reported to General Taylor with the somber news of the battle at Ebenezer Church. Taylor stated that, "Forrest appeared, horse and man covered with blood, and announced the enemy at his heels, and that I must move at once to escape capture." (Taylor, p. 219)

Taylor took the train to Meridian, Mississippi, while Forrest busily prepared for the defense of Selma. The 3.5 miles of earthworks surrounding Selma had been constructed months earlier, with the city protected by the Alabama River to the south, and fortified positions on the remaining sides. The alarm was sounded sending the local militia into the earthworks and the city of 15,000 evacuated all unnecessary citizens and materiel. Forrest bandaged his wounds, and the badly crippled Confederates awaited the inevitable arrival of the Union horde.

Again, Wilson received a stroke of luck as a knowledgeable local "citizen" provided him with details concerning the defensive positions surrounding the industrial city of Selma. At dawn on April 2, Long's division was again in the lead of the column with Upton's behind him and they both pressed forward to their attack positions. Wilson conducted a personal reconnaissance of the situation and validated the information learned from the informant. Late that afternoon, Wilson's Cavalry Corps descended upon their target.

The cavalry made repeated charges against the earthworks while the Union artillery engaged in a duel with the Confederate cannons. That evening as darkness enveloped the city, the Union cavalry punched through and swarmed through the city streets. Wilson reported that:

The troops, inspired by the wildest enthusiasm, swept everything before them and penetrated the city in all directions. I doubt if the history of this or any other war will show another instance in which a line of works as strongly constructed and as well defended as this by musketry and artillery has been stormed and carried by a single line of men without support. (OR 49-1, p. 360)

Selma provided Wilson the triumphant victory he had long desired for his cavalry corps, and he had beaten "that devil Forrest," twice, and in as many days. His complete victory and destruction of the fortified city netted the Union a strategic rail depot, 32 cannons, and 2,700 prisoners.

While Forrest's Cavalry Corps seemingly disappeared into the night, General Upton's troops pursued them until early the next morning, capturing four more cannons and several more prisoners.

The next day, General Winslow was placed in charge of destroying the city's industrial capabilities, while General Upton's division was deployed to pursue the Confederate cavalry. Upton's men spent the next several days riding through central Alabama, destroying bridges, industrial facilities, and searching for the enemy.

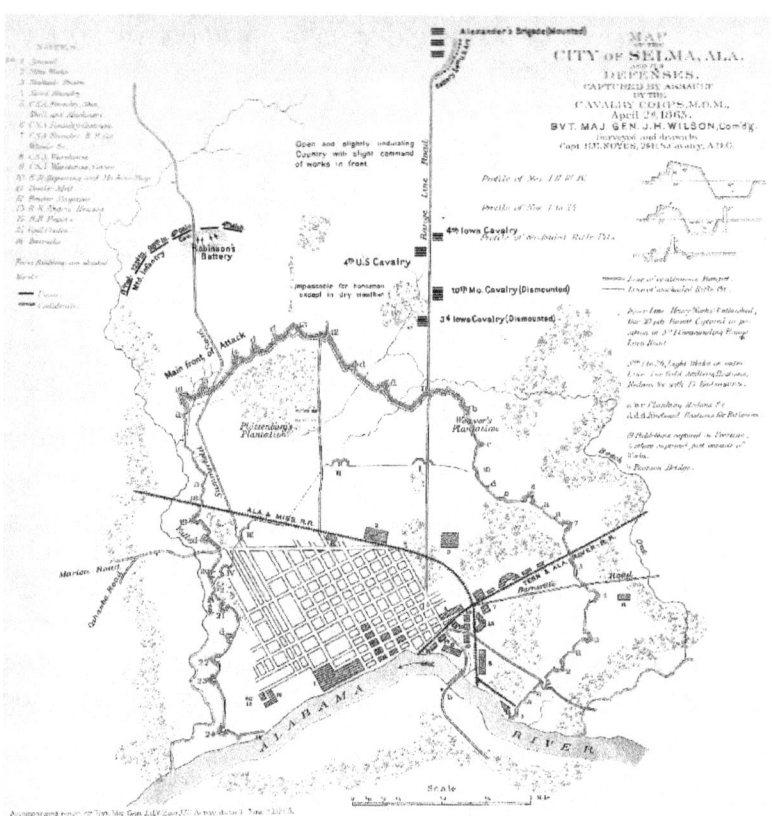
Battlefield Map of Selma

At Selma, Wilson rested from his recent battles and reflected upon his next course of action. General Canby had achieved victory at Mobile, and positive news was received that Grant was marching into Richmond. Wilson decided to move on to Montgomery and invade the state of Georgia.

With his enemy dispersed and badly beaten, Wilson spent the next several days trying to get a pontoon bridge established across the rain-swollen Alabama River so he could continue his mission. Wilson had anticipated this event and had brought with him a large contingent of pontoons and dedicated engineers to construct them.

Meanwhile, Forrest sent word to Wilson that he desired a meeting to discuss an exchange of prisoners. Forrest met with Wilson at a private residence in Cahaba, 10 miles below Selma. Forrest greeted his nemesis with, "Well General, you have beaten me badly, and for the first time I am compelled to make such an acknowledgement."

Wilson courteously responded that the victory was very costly, and "You put up a stout fight, but we were too many and too fast for you." Forrest replied with "Yes I did my best, but if I had your entire force in hand, it would not compensate us for the deadly blow you have inflicted upon our cause by the capture and destruction of Selma."

The two warriors dined together, and before the evening was through, Wilson recorded that "we were treating each other like old acquaintances, if not old friends." (Wilson, pp. 241-43). After dinner, the two generals went their separate ways.

General Forrest and remnants of his former units retreated to the east on the Burnsville Road, then swung north to Plantersville, coming around and behind the Union Cavalry. That night, Forrest's men killed dozens of the Union troops and then rode west toward Mississippi (he would later surrender there on May 9, 1865 at Gainesville, Alabama). Buford's men fled southeast from Selma and would continue to fight a long and bloody running fight against Wilson's men for the next several weeks. General Forrest, beaten and broken had killed his last man in battle, but for General Wilson there remained more bloody work on the road ahead.

When Wilson returned to Selma, his men were still working on their pontoon bridge. In their makeshift camp, Sergeant Benjamin Harrington a horse soldier in the 4th Iowa recorded in his diary, "Our success has been perfectly glorious. Thank God… Bathed in the Alabama River… Would love to hear from home … the pontoon was finished yesterday… then

the bridge broke... it was fixed... soon it broke again."
(Harrington, April 4-8)

Finally, after working night and day for one full week, Wilson's cavalrymen, all working together, managed to establish an 870-foot long pontoon bridge, and the corps crossed the rain swollen Alabama River. Unbeknownst to Wilson's Corps, while they struggled to cross the Alabama River, General Robert E. Lee surrendered his army at Appomattox, Virginia. It would be days before the news reached them.

On the morning of April 10, Wilson's Cavalry Corps was back on the march, and their next target was the former Confederate capital, Montgomery. While Wilson was rid of the threat from Forrest himself, General Abraham Buford continued to haunt the Union column.

With the roads muddied from the spring rains, Buford's men destroyed bridges, made obstacles, and engaged their enemy with hit and run assaults. While the distance from Selma to Montgomery is 50 miles, the journey took the Union cavalry two full days.

Wilson later reported that along their trip to Montgomery, the Confederate cavalry harassed them "all the way." While Buford worked so diligently delaying the enemy, his counterpart, General Daniel Adams was busily organizing the defense of Montgomery.

General Adams faced a daunting task, but he was prepared to die defending the capital city of Montgomery. While he was small of stature, Adams was a tenacious fighter. He had been wounded in battle three times, including the loss of an eye at the Battle of Shiloh. His brother Wirt was also a General, and he commanded a brigade in Forrest's cavalry. While General Daniel Adams was willing to give his all for "the cause" it was impossible to fill the trenches surrounding the city as he could only muster 1,800 men.

The mayor of Montgomery, Walter Coleman organized a committee and the nervous citizens urgently debated their predicament. News had reached Montgomery that Richmond had fallen, and there were rumors that Lee had surrendered to Grant. Mobile was under siege, and the bastions of Blakely and Spanish Fort had been lost. Elyton (Birmingham), Tuscaloosa, Montevallo, and Selma had all been sacked, and Forrest's Cavalry Corps was shattered.

Mayor Coleman appealed to General Adams to consider surrendering. (Rogers, p. 145-7) Adams refused to surrender the city, Governor Thomas Watts fled the capital, and the citizens of Montgomery destroyed 85,000 bales of cotton, countless gallons of whiskey, and set fire to the governor's stately mansion.

Meanwhile, General Taylor sent General Adams an urgent, abbreviated message to report to General Cobb and "unite his infantry with you for defense of Columbus, keeping Buford with cavalry to delay and annoy enemy. Re-enforce Buford with all cavalry you can, and strengthen Columbus works." (OR 49-2, p. 1239)

Just as the enemy was approaching the outskirts of the city, the last train from Montgomery was filled with the infantry and sent to Columbus. Just as the Union column approached from the west, Generals Buford and Adams escaped to the southeast on the Columbus Road, where they prepared more obstacles for Wilson's corps.

On April 12 at 7 a.m., and four years to the day after Fort Sumter fell to the Confederacy, the Union advance guard entered Montgomery. The mayor and councilmen met with General Edward McCook who formally received the surrender of their city. With his military band playing patriotic tunes, General Wilson had the "Stars and Stripes" hoisted upon the capitol dome.

Then, under the waving flag of the United States of America, Wilson's Cavalry Corps unfurled their unit colors, and triumphantly marched through the streets of Montgomery. The former Confederate citizens watched the parade and quietly accepted defeat.

While the city officials had hoped to avoid any destruction, all Confederate facilities were immediately destroyed, and provisions for the Union troops were forcibly procured from the citizens. The Confederate arsenal in Montgomery included an ammunition plant, quartermaster stores, and niter works, which were all thoroughly destroyed and burned.

A detachment of the Fourth Kentucky made a rapid march to the Coosa River (near Wetumpka, Alabama) where five Confederate steamboats and their cargoes were destroyed. (49-2, p. 405, Wilson, p. 250-53)

After reviewing the situation at Montgomery, General Wilson contacted General Thomas with the following message:

Buford and Adams have fled again with about 1,500 men toward Columbus… I shall move in the same direction early tomorrow. I had determined on this course after careful deliberation and upon conviction that I shall best accomplish what is expected of me by you and Grant. I am sure Canby will experience no serious difficulty taking Mobile, subjugating entire State, and breaking up all rebel force between Sherman and the Mississippi River. The people say he now has Mobile.

> *The destruction of Selma and defeating of Forrest have deranged rebel plans. Fall of Richmond and defeat of Lee have deprived rebels in this section of their last hope.*
>
> *If I can now destroy arsenals and supplies at Columbus and divide their army in the southwest, they must disintegrate for lack of munitions. There is no force to resist mine, and I see no reasonable ground for fearing failure. My command is in magnificent condition; every man is splendidly mounted, plenty of forage and supplies of all kinds.* (49-2, p. 344)

With the Confederate cavalry forces of General Forrest broken and in disarray, the domineering Union raiders were free to continue into Georgia. General Wilson now focused his hugely successful raid on Columbus, Georgia.

According to General Wilson, Columbus, Georgia was both the "key" and "door to Georgia," but this city was also a Confederate prize unto itself as its vast resources and industrial capacity were second only to Richmond. (OR, ser. 1, Vol. XLIX, Pt. 1, P. 364; Wilson, Old Flag, 11:265)

An assault on Columbus required the capture of a bridge with which Wilson could rapidly cross his corps. Within striking distance were several bridges that crossed the deep divide between Alabama and Georgia. While Wilson's corps had a pontoon unit, they required too much time to construct their equipment, and with the Confederate cavalry harassing their every move, the task would consume time and present great hazards.

The Union cavalry could swing wide to the north or south of Columbus, but this maneuver would also create an undesirable delay. Wilson had achieved great success with his raid thus far, and his rapid pace had served him well. Militarily, "surprise and violence of action" are the two key components of a successful raid and Wilson repeatedly harnessed them. He weighed his options and chose to "blitz" the objective.

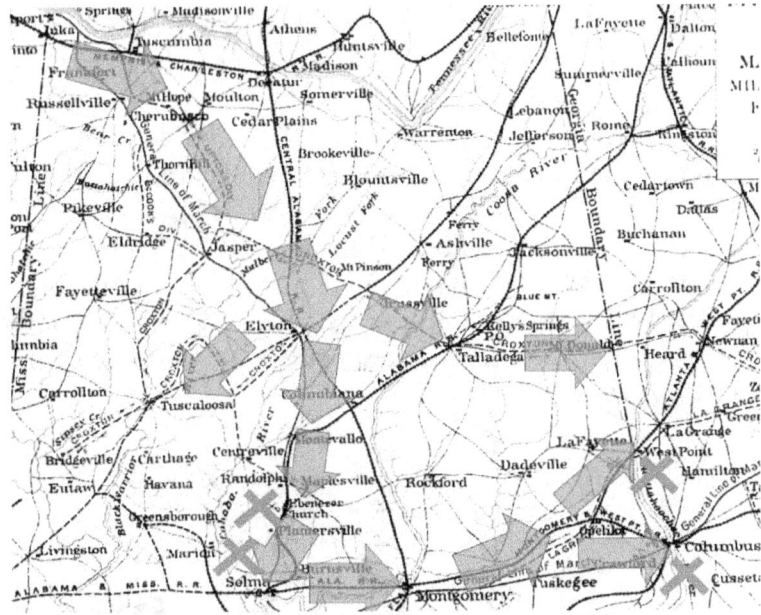
Route Map of Wilson's Raid across Alabama

General Wilson's first objective in attacking Columbus, Georgia was to secure a bridge crossing the Chattahoochee River. In case the Confederate defenders destroyed the bridges at Columbus, Wilson hoped to secure a bridge crossing at West Point, Georgia. He accordingly divided his forces to attack Columbus and West Point on the same day.

General Wilson was confident that he could take at least one bridge, and that was all he needed. The brigade commanded by Colonel LaGrange of McCook's Division was assigned the mission of securing the bridge at West Point, Georgia (35 miles northwest of Columbus), while General Emory Upton, commanding the Fourth Division was given the main target, Columbus, Georgia. Columbus had two wagon bridges, one foot bridge, and a railroad bridge.

Thus, between the two targets, West Point and Columbus, there were a total of five bridges and surely one could be taken intact. Wilson assigned General McCook to complete the destruction of Confederate equipment and supplies in Montgomery and then follow the main column to Columbus.

General Upton was ordered to provide a detachment to remain in Montgomery as a Provost Guard, which he assigned to Major Kirkendall with six companies of the Third Iowa, E, F, G, H, L, and M. Colonel Robert Minty's Second Division (General Eli Long was wounded at Selma) was assigned to the rear of the main column, which was led by Upton at the head of his division. With his plan secured, Wilson's army prepared for its execution.

Colonel Robert H. G. Minty

Before the sun reached the horizon on the morning of April 14, the bugler shattered the silence with the audible signal of "Boots and Saddles," putting the Union cavalry back into motion. As the men readied their mounts, and packed their gear, news spread through the ranks of a great Union victory in Virginia. After four long years of war, Union morale was reaching its zenith and the cavalrymen set forth to take Columbus.

According to Sergeant Harrington of the 4th Iowa, Wilson's Corps, "Commenced marching early…Richmond is ours, 'twas taken in 3 days terrific fighting. This the Rebs admit… God be praised." (Harrington)

Another of Wilson's officers, Captain Charlie Hinricks of the 10th Missouri, wrote that he was "more than ever satisfied that at an early day the Confederate bubble would burst." (Hinricks April 13) Still, there was plenty of fighting on the road ahead as the "Confederate bubble" had not yet burst. According to Sergeant James Larson of the 4th Iowa Cavalry:

> *We moved on Columbus, Ga. on the Chattahoochee River. From then on, the skirmishing again became constant and daily. The enemy's cavalry hovered on our flanks and opposed our advance, making their stands at all naturally strong positions.* (Larson, p. 301)

The harassing attacks of the Confederate cavalry affected every unit in the Union column as the road from Montgomery to Columbus ran through dense forests that provided excellent concealment for Buford's cavalrymen to emplace obstacles and ambush the advancing Union column.

The threat of a sudden and violent death while on the march unnerved the Union troopers. The persistent fighting took its toll on the Confederate cavalry as well, and according to Buford's men, "blood was shed on every mile to Columbus." Captain Charlie Hinricks of the 10th Missouri recorded that "at every crook and corner on the road, the enemy would fortify and barricade with rails, give us a volley, and fall back." (Hinricks, 14)

According to Wilson's adjutant, E.N. Gilpin, during the skirmishing on the 14th, "we had five men killed today." (Gilpin, 651) On the evening of April 14, unbeknownst to the Union cavalrymen, their Commander in Chief, President Abraham Lincoln was shot by an assassin while attending a play. By the next morning, the president was dead, but the news would not reach Wilson's cavalry for several days to come.

The 15th of April brought hard thunderstorms, which complemented the Confederate efforts to delay the marauding Union troops. Mud made the roads a treacherous morass, and Buford's cavalry incessantly harassed their Union foe. Again, Sergeant Larson noted, "Those gentlemen in grey were always showing a most unneighborly [sic] disregard for our comfort." (Larson, p. 208)

While these minor skirmishes caused several casualties and a lot of work for the Union cavalrymen, the main column was never really impeded as the scouts cleared the way for the main column behind them. Despite their many hardships and obstacles, the Union cavalry made good time and the evening of April 15th they arrived and encamped at Tuskegee, Alabama.

This town also surrendered without a fight, but the local press provided a warning for what awaited the Union troopers in Columbus. Charlie Hinricks, recorded, "We encamped within 25 miles of our next objective point- which is Columbus, Georgia." In a Columbus paper dated the 14th, he read that the rebels are "determined to give battle. I will lay down and rest- I shall need it tomorrow." (Hinricks, April 15).

According to Captain William Scott, Adjutant of the 4th Iowa, "The next morning, the 16th, as the column moved out, it was inspected and prepared for conflict with all the care and rigor that had been observed on the morning of Selma, and every soul was filled with the hopes and fears of another hazardous battle." (Scott, p.482)

The column moved out early that morning, but slowly, as the road was in terrible condition and muddy from the recent rains. The terrain was swampy, dense with vegetation, and the advance party had to corduroy the muddied route in many places. Again, Buford's men attacked the invaders; however, General Upton dispatched a squadron of cavalry led by Captain J.A.O. Yeoman, of Company A, 1st Ohio Cavalry Regiment to press forward and secure two bridges that were being held by the Confederates just east of Crawford, Alabama.

These two bridges were small, but critical to the continued march of the main column as the bridges were in a large swampy marsh that surrounded the Little Uchee Creek.

General Upton wanted both bridges secured for safe passage, as the column was to split after passing through Crawford. The town of Crawford lies just 14.5 miles west of Columbus, and provides several routes by which Columbus can be reached. Colonel Minty in command of General Long's Division would take the main road (now Highway 80) straight through to Columbus, while Upton's Fourth Division would take the lower road (now State Routes 169 and 28/ County Road 41 and 27).

Captain Morse split his force into two elements with each assigned a bridge to secure, and off they raced on their assigned missions. While the Confederates attempted to defend the bridges, ultimately, they set fire to them and retreated towards Columbus.

The Union cavalrymen managed to extinguish the flames, and secure passage for the main column, which was advancing at a steady pace, one hour behind them. The creek that passed beneath two small, but important bridges was the final obstacle, which could have created a considerable delay for the main column. After passing through Crawford, the main column diverged along their planned separate routes and they continued their respective marches towards their target, Columbus, Georgia. (Yeoman, 222-23)

CHAPTER 3
COLUMBUS, THE CITY

The City of Columbus, Georgia was established in 1828 as a trading town on the Chattahoochee River. Situated in the southern half of Georgia and directly upon the western boundary of the state, Columbus also serves as the county seat of Muscogee County. Looking westward across the river from Columbus is Phenix City (formerly known as Girard), the seat of Russell County, Alabama.

While most of the land this far south is generally flat, this area lies on the fall line where the Appalachian Mountains terminate in their southernmost extension. Thus, the terrain is divided between two geographical regions, the Piedmont Plateau and the Coastal Plain. The elevation varies from 180 to 650 feet, with the dominant heights on the Alabama side of the river.

These heights consist of a group of hills averaging 450 feet in elevation that form a small valley, in which lies the municipality of Phenix City (Girard). Mill Creek (now known as Holland Creek) drains the surrounding high ground and flows eastward through the city, and empties into the Chattahoochee River, just above the Dillingham Street Bridge, and directly across the river from Columbus.

This region lies in the heart of the Chattahoochee Valley, and the flow of drainage is southward through the valley, which ultimately empties into the Gulf of Mexico, several hundred miles below. The region's soil consists primarily of Norfolk sandy loam and clay depending upon the elevation and proximity to natural drainage. (Muscogee stats) Its forests contain a mix of both deciduous hardwoods and evergreens, but once settled, much of the land was quickly cleared and planted.

The climate is generally mild throughout the year except for summer, which is long, hot, and humid.

Temperatures average 48 degrees Fahrenheit in the winter and 81 degrees Fahrenheit in the summer. Rainfall averages 49 inches annually, and the average growing season for agricultural products is eight months.

During the decades of 1830 through 1860, "King Cotton" dominated the region and Columbus was established as a transshipment point for the valuable crop. Steamboats laden with cotton plied the Chattahoochee River and the region flourished.

Westward expansion and land availability provided the area with a population that quickly swelled, and Columbus experienced tremendous growth. Numerous plantations were established throughout the region, and with Columbus serving as their hub; industrial growth was phenomenal.

While the town of Girard, Alabama also grew, most of its citizens worked across the river in Columbus. People flocked to the Columbus area for its many opportunities. Mills exploited the natural power of the river. Railways and paddlewheel steamers carried goods to and from the city.

By 1860, Muscogee County had a population of 17,000 inhabitants. Columbus was the third largest city in Georgia, "with a total population of 9,621, of which 5,933 were white, 3,547 slaves, and 141 free blacks. Within Muscogee County, 40% of the 1,972 families owned slaves." (Dameron, 89) While much of the regional populace engaged in farming, the citizens of Columbus and neighboring Girard were predominantly employed in local factories. The ante-bellum community was filled with stately homes and progressive growth. Columbus was proud of its manufacturing capabilities and it was commonly referred to as the "Lowell [Massachusetts] of the South."

Roads for stage, wagon and rail were quickly established and Columbus became a regional hub. In 1853, the Muscogee Railroad connected with the Southwestern Railroad at Butler, Georgia, which provided a direct link to Macon, and thence to the Port City of Savannah. Thus, connected to Macon via rail and telegraph to the east, southward via the river to the Gulf, Columbus also extended her growth westward.

Girard, Alabama was the home of an acclaimed bridge builder, a free black man named Horace King. This talented engineer built several covered wagon bridges that spanned the Chattahoochee connecting roads between Alabama and Georgia. (Coulter, 116)

The people of Girard, Alabama complemented the growth of Columbus by extending roads and rails to the west. Completed in 1856, the "Opelika Branch" of the Montgomery and West Point Railroad (also known as the Columbus & Western R.R.) connected the Columbus/ Girard terminal and all points eastward to Opelika, whose depot connected with all points west. (Martin 56-7)

Another direct line, The Mobile and Girard Railroad was begun in 1850 and this line was designed to provide a direct rail link to Mobile. While this line was incomplete during the Civil War, its 38 miles of track provided important services for local planters hauling cotton to market. (Standard, 15)

The little town of Girard and the community of Brownville eventually grew into what is known today as Phenix City, Alabama, and its growth continues to support the community with resources and opportunities on the west bank of the river.

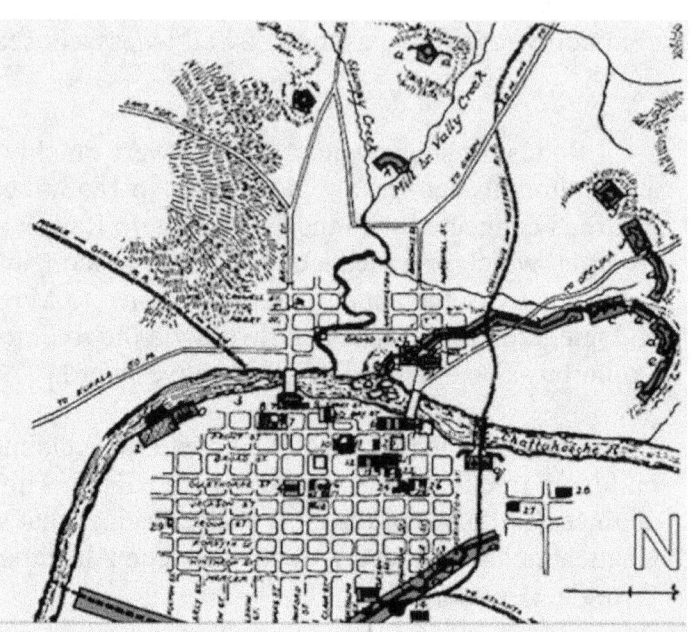

DEFENSE WORKS

a- Rifle pits
b- Fort, 30 yards, square, 4 guns
c- Lunette- 2 guns
d- Heavy fort, 200 yards long, 4 guns
e- Lunette, 1 gun
f- Heavy fort, 150 yards long, 3 guns
g- Heavy fort, 200 yards long, 4 guns
h- Heavy fort, 3 guns, steep hill
i- Fort, 30 yards, square, 4 guns
j- Heavy fort, 150 yards long, 3 guns
k- Heavy fort, 3 guns, steep hill
l- Battery, 6 guns, supporting b
m- Battery, 6 guns, supporting b
n- Battery, 4 guns covering lower bridge
o- Battery, 4 guns covering Navy Yard
p- Battery, 2 guns covering upper bridge
q- Battery, 4 guns covering RR bridge

CSA INDUSTRY

1- Ironclad, CSS Jackson
2- Navy Yard
3- Steamboat landing
4- Court House
5- Lunette, 1 gun
6- Railroad Station
7- Eagle & Phoenix Cotton Mill
8- Columbus Ironworks
9- Rolling Mills
10- Commissary Department
11- Confederate Hospital
12- CSA Government Offices
13- CSA Government Shops
14- CSA Laboratory
15- Murray's Gun Shops
16- Gunther's Wagon Shop
17- Shoe and Button Shops
18- 21 Hospitals
22- Sword Factory
23- Shoe and Harness Shops
24- Gun Factory
25- Haiman's Sword and Bayonet Factory
26- Haiman's Pistol Factory
27- Wheelbarrow and Button Shop
28- Glue and Oil cloth Factory
29- Oil Cloth Factory
30- Armory

While Columbus has always dominated the valley in industrial growth and population, her western neighbors in Alabama contribute greatly to the success of the region. With a shared heritage and an entwined future, these twin cities on the Chattahoochee continue their unique association.

A uniquely interesting occurrence that parallels the growth of the region concerns its military heritage. The Creek Indians had established a village known to them as Coweta Town on the banks of the Chattahoochee River many years before it became Columbus. When the Creeks ceded their lands to the State of Georgia in 1825, the Indians simply moved across the Chattahoochee into Alabama, which they originally owned too.

This area was known as the Creek Nation. In fact, a preponderance of geographic names throughout the region are derived from these Indians. The Muscogee were the predominant tribe of Creeks, and a very proud people. While they generally remained on friendly terms with their white neighbors, they did not willingly comply with the demands of the U.S. government and the encroaching settlers.

In accordance with the Treaty of Cusseta in 1832, the heads of the Creek families were given individual tracts of real property in the Creek Nation, but were required to sell or dispose of their lands by 1836, when they were to emigrate to the West.

As the time for their departure approached, some of the Creeks waged war against the settlers who were intent on ridding the Indians from the region. The local militia joined with the U.S. Army to fight the Indians, who were ultimately defeated. The Indians were then forcibly removed from Alabama and Georgia along the infamous "Trail of Tears" to reservations in Arkansas and Oklahoma.

Thus, when Columbus was established in 1828, the proximity of hostile Indians necessitated an organized militia. This in turn created a tradition in the Columbus area wherein military organizations were not just widely supported, but revered as a chivalric duty of manhood.

During the Creek War in 1836 and the Mexican War in 1846, Columbus eagerly provided troops and honored their services to their local community, as well as their state and nation. Most of the city's prominent citizens were a part of these military organizations and their participation therein interlaced the local community in social, business, and military affairs.

Throughout the ante-bellum period, Columbus maintained at least three military companies that performed regular drills and attended state musters. (Martin, 94, 104) These military traditions and acts of preparedness would prove very beneficial to the region during the Civil War.

While the locals extolled the virtues of military preparedness, they equally pursued religious and leisure activities. By 1860, Columbus had 13 churches, several horse racetracks, a theatrical center, numerous clubs and societies, as well as several local fairs. Columbus was filled with an active social scene complete with gala events such as balls, fundraisers, and political rallies.

While churches played an important role in the community, whiskey flowed freely and a plethora of bars, with names such as *"The Smile,"* and *"The Pleasant Hour,"* were scattered throughout the downtown area. In professional services, the city had 43 Lawyers, 19 Doctors, 6 Dentists, 6 bank agents, 6 insurance agents, and a multitude of service providers.

Downtown, Columbus was filled with several bookstores, music stores, drug stores, dressmakers, hat makers, shoe stores, restaurants, hotels, cigar shops, lumberyards, and furniture stores. Three local newspapers: *Columbus Daily Sun, Times, and Enquirer* provided the region with excellent press reports.

The newspapers were filled not only with the myriad reports of state and national news, but also the excitement of local industrial modernization. In manufacturing, Columbus had a total of four major textile mills, three large gristmills, a large cotton gin, a steam engine builder, three planing mills, two carriage dealers, an iron works, and a machinist firm.

Early in 1860, the Eagle Manufacturing Company purchased the Howard Factory, and the local newspaper reported, "We understand that these factories run 10,000 cotton and 1,300 woolen spindles; that they have 282 looms weaving cotton and 1,000 pounds of wool per day, and employ 500 hands, at an expense of $240 per day for their labor."

On a smaller scale, one of the other textile mills "the Grant Factory produced 480,000 yards of osnaburg, 80,000 yards of kerseys, 78,000 pounds of yarn, and 6,000 pounds of rope." *Columbus (Georgia) Enquirer*, April 10, 1860; Martin, Columbus, Georgia, From Its Selection as a "Trading Town" in 1827, To Its Partial Destruction by Wilson's Raid in 1865. Volume 1, p. 118; *Daily Sun* (Columbus), March 16, 1861; April 8, 1863; Standard, *Columbus, Georgia in the Confederacy: The Social and Industrial Life of the Chattahoochee River Port*, pp. 28-9, (a detailed narrative of Georgia's war time industrial production).

Amongst the most important firms in Columbus, the Eagle Mill provided the local economy with its greatest resource. It served as one of the South's largest cotton mills as well as its most progressive. As early as 1838, females worked in the Columbus cotton mills and in 1854 an advertisement for the

Eagle Mill, enticed prospective female employees with an offer of "$10.00 per month, plus housing, and medical care."

The proprietor of the Eagle Mill was William Young, a New Yorker that moved to Georgia at the age of seventeen and adopted Georgia as his home. He realized the potential of textile manufacturing in Columbus as there were several mills already in service in Columbus, but Young capitalized on the resources available and in 1860, he purchased the Howard Manufacturing Company and quickly turned his company into a highly successful venture.

With the water power of the Chattahoochee, the vast quantities of cotton, and a large pool of laborers to make textiles, Young had the Eagle Mills spinning cotton as never before. He also served as the President of the Bank of Columbus and Young's skills in business were only surpassed by his benevolence to his adopted community. His concern for the local citizenry was genuine and in turn, the people of Columbus respected him as a good man and a pillar of the community.

Despite the tremendous growth and livelihoods being enjoyed in the Columbus area, everything soon changed. In 1860, Abraham Lincoln was elected as the 16th president of the United States, and as threatened, the people of the South refused to accept him.

Debates concerning secession and the formation of a new country dominated everything. Many of the leading citizens of Columbus entered the political scene. Among these men were Henry Benning, a successful attorney, and former State Supreme Court Justice.

Benning had lived in Columbus for 26 years, he had served in the local militia during the Creek War, and he was well respected throughout Georgia. He was a classmate of

Georgia's leading statesmen, Alexander Stephens and Howell Cobb, but he was more outspoken and determined than they to secede.

Benning joined with his best friends, Martin J. Crawford, also of Columbus, and William Yancey, an experienced statesman of Alabama, to win the case for secession. Together they traveled throughout the south, represented their communities at debates, and ultimately served as catalysts in the creation of the Confederacy. Then, in a flurry of events, the nation plunged into a rapid succession of events that ultimately culminated in the abyss of war.

In January of 1861, Georgia signed a declaration of secession. The Confederate States of America was formed in March, and shortly thereafter, on April 12, 1861, the Confederates fired on Fort Sumter initiating the American Civil War. Throughout these momentous occasions, Georgians reached a fevered pitch of excitement.

In Columbus, "the most brilliant display ever witnessed," filled the streets with military parades, ladies in their best gowns, and torchlight processions. Above the streets, the air was full of fireworks, patriotic banners, the clanging of church bells, cannon fire, marching bands, shouts of cheer and enthusiastic speeches.

Volunteers flocked to join the military companies, forcing the establishment of new units. In addition to the ever-vigilant Columbus Guards, by February of 1861, Columbus also had the Southern Guard; City Light Guards; Georgia Grays; and the Muscogee Mounted Rangers. Equipped with uniforms manufactured by the Eagle Mills, and outfitted with arms, these units formed the initial vanguard of the city's voluntary spirit. Under the watchful eye of Brigadier General Paul Semmes of the Georgia State Militia, these companies drilled and practiced in the city commons.

Recruitment was robust, as the entire community was filled with martial spirit and enthusiastic support. Additionally, the Columbus manufacturing firms and local entrepreneurs converted their production efforts to meet the needs of the military. Local entrepreneurial merchants and textile firms won contracts to supply: tubs; buckets; tents; shoes; uniforms, and military buttons embossed with the Georgia coat of arms. (Dameron, p 110-15; Mosocco, *The Chronological Tracking of the American Civil War per the Official Records of the War of the Rebellion*, p. 6 (excellent reference source for accurate dates of events during the Civil War); The Daily Times (Columbus), January 13, 1861; Worsley, p. 272, *Columbus on the Chattahoochee*; Ibid., February 21, 1861; The Daily Sun (Columbus), April 17, 1861; Worsley, Columbus on the Chattahoochee, p. 272; *Columbus* (Georgia) *Enquirer*, January 2, 3, 1861, March 20, 1862; The Daily Sun, January 1 and 31, 1861, April 1, 1861.

In 1861, the Chattahoochee Valley was filled with the initial "atmosphere of the bayonet," which permeated the Confederacy. Filled with patriotic devotion and a warrior spirit, citizens and soldiers alike rushed to the aid of their new nation. In the *Columbus Enquirer*, June 20, 1861, it was noted that the city of Columbus had provided ten companies of soldiers, "enough to make a full regiment," and "our citizens have contributed $1200 for each company." It was also noted that these figures did not include the private donations of food, clothing, and equipment necessary to outfit these units.

Yet, concerning these numerous donations from the citizens of Columbus, it was also recorded that, "Surely, she has done all that could be expected of her, unless the emergency was such a one as to require every citizen of the South to become a soldier."

During the next four years, nearly everyone in Columbus would be required to provide some form of service to the Confederacy. While Southerners were prepared to fight, they were naive concerning the immense costs involved with war, which included both blood and money. (Dameron 118; *Columbus Daily Enquirer*, June 20, 1861)

The common men and women of the Chattahoochee Valley contributed both blood and money to the Confederacy. Their voluntary spirit and contributions to "the cause" were unmatched throughout the South.

Between May 1861 and February of 1862, the city of Columbus contributed sixteen Infantry, and two Artillery companies to the ranks of the Confederate Army. Each company contained an average of ninety men and the approximate number of white males living in the city was just 3,500.

Thus, conservative estimates reveal that nearly one-half of the available population of Columbus carried a weapon into war. Likewise, across the river in Girard, the Alabamians eagerly filled the ranks of their state units as well.

For the first few years of the war, the central region of both Alabama and Georgia remained unthreatened, but in 1863 the Union army began pressing into the area. The primary Union thrust was during the fight for Chickamauga, but another little-known Union incursion caught the attention of the Columbus area. During April and May of 1863, Colonel Abel Streight led a Union brigade in a daring raid across the state of Alabama and into Georgia.

General Nathan Bedford Forrest and his famous cavalrymen chased the Union raiders across central Alabama and cut them down before they could do any damage in Georgia. While the raid was a bold maneuver, it achieved little and ultimately the Union raiders were all killed or sent to Confederate prisons. (Series 1, Vol. 23, part I, p. 283-85)

Columbus got involved when Forrest and his men passed through the city and they were hailed as heroes for destroying the "Yankee threat" to hearth and home. The Ladies of Columbus were so overcome with adulation that they pooled their resources and bestowed a gift of a magnificent horse upon General Forrest for his courageous actions. From that day forward, General Forrest proudly rode this handsome steed known as "King Phillip."

Described as a "great a horse as Forrest was a soldier," King Phillip had a pale white coat and a shiny black mane and tail. (CV Vol. XXXV, April 1927, p. 138) The pair were inseparable through the remainder of the war and Forrest cherished his fine stallion. The ladies of Columbus would be repaid in devoted service and relentless pursuit of any Union raiders that threatened Alabama and Georgia.

During the first few years of war, the Confederacy won several impressive victories, but it also suffered a multitude of casualties.

One of the wounded veterans that returned home was the ex-Mayor of Columbus, Francis Wilkins. Upon his return home, he encouraged everyone that could to volunteer for the war, but the Confederacy could not manufacture soldiers. The sons of the South were a limited commodity, and except for those men who were either, too old, invalids, or too young to fight, there simply were not many to be had.

Nonetheless, Wilkins returned to his duties as Mayor and he focused his efforts on the defense of his home, raising a non-conscript "invalid corps," which provided a reserve composite force of three hundred men organized for home defense.

While the locals could not imagine an invasion of their homeland, realists knew that the more industrial and more populous north was an ominous foe. As the war progressed, the strength of industry in Columbus matured as well. Private firms eagerly converted their efforts to produce wartime materiel, and the Confederacy rewarded them with a bounty of work.

By April of 1863, both Columbus and Girard had contributed their full complement of human resources to the production of war materiel, and nearby residents flocked to the twin-cities in search of jobs. Despite the thousands of soldiers that had left the area, the population of Columbus increased to 15,000. Thousands of people flocked to the area to work in the factories, and to maximize labor and meet their production needs, young boys, girls, and Negroes were thrust into work.

With the ability of the region to ship products via rail, steamboat, and wagon, a bounty of Confederate contracts engaged nearly all firms, large and small. Several firms merely changed their customer base to focus on the needs of the Confederacy.

For example, the Rock Island Paper Mill used cotton waste from the local textile mills and supplied the Confederacy with tons of paper products throughout the war.

Another firm, the Muscogee Iron Works ceased forging farm equipment and began casting cannon barrels. Three other large mills produced 300 barrels of flour and cornmeal per day, which was primarily, diverted from civilian grocery stores to quartermaster supply wagons.

The Confederate Army needed lots of wagons too, thus Barringer and Martin, former-building contractors converted their sawmills and lumberyards into a wagon and gun carriage enterprise.

Textile mills, which consisted of the Eagle Mills, Clapp's Factory, and the Grant Factory, which prior to the war wove cotton and woolen cloth, converted their 152 looms and 9,800 spindles to make tents, uniforms, knapsacks, and oil cloth. (Standard, 28-9) A local tailoring firm, S. Rothchild and Brother worked frantically to fashion the gray tweed provided by the mills into regulation Confederate uniforms.

Nearby, another businessman, S.M. Sappington, a Broad Street grocer changed his store into a shoe factory, and the demand always outweighed his capability of supply. This facility produced thousands of leather shoes for soldiers who quickly wore them out. (Standard 32 and 36)

To harness the full potential of the region's manufacturing capabilities, the Confederate Quartermaster Department needed a full-time officer to manage the myriad industries, coordinate their efforts, organize transportation, and ensure a constant flow of war materiel. Frank W. Dillard was immediately recognized as the perfect man for the job.

Prior to the war, Dillard was a local merchant and co-owner of Dillard, Powell, and Company and he was an avid secessionist. He also served as a senior officer in the "Knights of the Golden Circle," which transitioned into the "Southern Guard" as war loomed on the horizon.

The "Southern Guard" was the umbrella organization for all military companies that formed in Columbus, and it was divided into two parts, military and civilian. General Paul J. Semmes of the Georgia Militia led the military operations while Dillard served as the senior officer of the civilian branch. Dillard's duties included the supervisory functions of clothing and equipping local military units as they prepared "to protect Georgia or any other Southern state."

In July of 1861, Dillard was appointed by the Confederate government as a Captain and placed in command of the Quartermaster responsibilities in Columbus.

Dillard's first act was to contract the production of 20,000 Confederate gray uniforms. With the local textile mills churning out rolls of gray cloth, and buttons being stamped out in mass quantities, Dillard contracted with local tailors to complete the uniforms and get them to the field. (*Sun*, June 10, 1861)

Unbeknownst to the Confederate authorities, Dillard had enlisted voluntary female seamstresses to complete the initial order. Dillard's success resulted in increasingly larger orders and a promotion to major on October 7, 1862.

Along with his promotion, Dillard received more complex duties as he was assigned to coordinate all contracting and manufacturing as the Post Quartermaster of the Columbus Depot. Major Dillard was soon filling trains laden with Columbus Depot uniforms. Over one million garments were manufactured in Columbus, Georgia for the Confederate cause.

From all these uniform items made in Columbus, the Columbus Depot Jacket was very popular amongst soldiers of the Confederacy, and it remains toady a favorite of military collectors. Dillard enlisted the assistance rendered by the ladies of Columbus who shocked Dillard one morning as he went to work and found, "women and girls waiting their turn to receive goods, cut to hand, to make for army use."

Dillard established a Confederate sewing center in Columbus to maintain high rates of uniform production and the official records reflect that he employed 2,000 women. Initially, these women were paid $1.00 per complete uniform, which in 1861 could be used to purchase two pounds of bacon.

Throughout the war, thousands upon thousands of uniforms were made in Columbus and as inflation rose, so too did the pay for the female seamstresses, and by the war's end, the ladies made $3.50 per uniform. Major Dillard was involved with the manufacturing of virtually everything a soldier needed, to include his shoes.

Shoes were always in great demand as the Confederate army literally marched them right off their feet. For Infantrymen, a good pair of shoes lasted an average of 3 months and the Confederate Quartermaster shops never fulfilled their demand. Still, Major Dillard did everything possible to increase production and he organized the largest shoemaking center in the Confederacy at Columbus.

Many free blacks worked in the government shops and as evidenced in the local press, Major Dillard solicited their help in the "Want Ads" with, "ONE HUNDRED NEGRO SHOEMAKERS to work in the Government Shop. Call at once and help me to shoe the army. Liberal prices will be paid." (CE August 12, 1864)

The Confederate Nail factory in Girard provided nails for these shoes, and hides were procured locally until leather became so scarce that Dillard had to look elsewhere. The Confederacy then authorized Major Dillard to procure hides in North Carolina and Tennessee to meet the demands of his shops and the enterprising officer continuously exceeded the expectations of his superiors.

Teaching unskilled craftsmen, the art of shoemaking and maintaining a steady supply of leather was a great challenge, yet Dillard employed just over 500 people in the production of shoes and at its peak capacity, the Columbus Depot made 5,000 pairs a week. (Standard 38-9) By the spring of 1863, the Columbus Depot was second only to Richmond in its Quartermaster production capabilities.

The smaller operations presented the greatest challenges and risky ventures for entrepreneurs. These businesses required re-tooling as enterprising shop owners converted their pre-war efforts into badly needed items such as wooden canteens, cartridge boxes, saddles, and harnesses. Jewelers and tinsmiths converted their products into military buttons, bayonets, and swords.

Rifles were also made in Columbus and two gunsmiths, J.P. Murray and John D. Gray ran small shops that produced rifles primarily for Alabama units, as Georgia procured the bulk of her rifles from larger arsenals. Still, these firms were in complete production by 1862 with Gray producing "Mississippi" Rifles (Enfield pattern) while J.P. Murray manufactured "Dickson-Nelson" Rifles and "Davis-Bozeman" Rifles. A lack of detailed information about these firms and the scarcity of lockplate stamped "Columbus Armory"

Rifles have generated a great deal of misinformation amongst collectors of rare firearms. Close examination of rifles stamped "Columbus Armory" and "J.P. Murray" reveals readily discernable characteristics, which provides collectors with the ability to identify many unmarked rifles as products of the Columbus Armory. The Columbus Armory made rifles and carbines, which were shipped primarily to Tennessee and Alabama.

Mr. Gray also operated a facility in Graysville, Georgia, which manufactured stocks for his rifles as well as military pikes and knives. At least 1,445 pikes were made and sold to the State of Georgia, and these are very rare items as their utility ended during the Civil War.

While Gray initially produced pikes, knives and rifles, in 1863, he converted his manufacturing operations to more mundane products whose demand warranted immediate attention.

Thus, the Columbus Armory ceased producing rifles and focused on products such as kettles, frying pans, chains, shovels, picks, and axes.

These products were made in the thousands and delivered to Major Dillard at the Columbus Depot for shipment to the army. Another weapons manufacturer in Columbus was Mr. A. H. DeWitt, but his attempts to manufacture swords for the army failed, as his products did not meet the stringent needs of the Confederate Ordnance Department.

Beginning a parallel effort in sword production like the DeWitt operation was another Columbus firm, which was owned and operated by the Haiman brothers. Louis Haiman, a Prussian tinsmith by trade, had immigrated to the United States with his brother, Elijah and they both settled in Columbus prior to the war.

While their business was small, they took great pride in their craft, and they capitalized on the opportunity to assist the Confederacy. The Haiman brothers established a shop on Dillingham and Short Streets where they produced brass buckles, bayonet scabbards, camp stoves, cavalry sabers, and swords.

As with other war industries, the demand far outweighed their productions. An estimated 8,000 swords were delivered to the Confederacy by the Haiman Brothers.

The Haiman Brothers produced quality products and their swords are worth more today than most Confederate swords. The first sword they made was presented to Colonel Peyton Colquitt and it is indeed a beautiful weapon. Exquisitely engraved and inlaid with gold the Haiman brothers produced the finest swords in the Confederacy.

Today, Haiman swords sell for thousands of dollars, but when they were originally made, they sold for far less. During the war, the Haiman brothers ran the following advertisement in the local press:

> *Swords! Swords! The best quality of swords are now made and for sale at the "Confederate States Sword Factory", Columbus, Ga. by L. Haiman & Bro. who have large contracts for the Confederate Government.*
>
> *They will furnish officers swords with belt for $25 or for $22 where as many as four are ordered in one lot. Every sword is tested according to the rules laid down in the manual of War. (Columbus Times,* November 17, 1861)

"L. Haiman & Bro." proudly affixed their name to their wares. Confederate officers highly prized these quality tools of war and Clanton's Alabama Cavalry Regiment was completely outfitted with Haiman sabers. (Albaugh 57)

At its zenith, the Haiman sword factory employed 500 workers, and when quality steel ran short, Elijah Haiman went to Europe to ensure it was shipped through the blockade. Ultimately, the Haiman brothers achieved the largest and most productive sword factory of the Confederacy.

In 1864, the Haiman brothers added several pistol experts to their staff, and they made a CSA version of Colt pistols as well as swords. While, their pistols never achieved the production rates of their swords, nonetheless the Haiman brothers and other arms manufacturing firms of Columbus contributed immensely to the war effort.

Another successful firm was the Columbus Arsenal, which produced ammunition for rifles, pistols, and cannons. The Columbus Arsenal achieved an average daily production rate of "10,000 rounds of small arms ammunition and 75 to 100 rounds of artillery shells."

In fact, Columbus served as the Confederacy's fourth largest producer of ammunition, easily surpassing the rates achieved in Charleston, Macon and Selma.

In July of 1864 as Sherman threatened Atlanta, the Confederate authorities shipped its entire Atlanta Arsenal operation to Columbus.

While the additional machinery was soon "on-line," the operation was hampered due to a shortage of supplies. The Confederacy lacked sufficient quantities of lead for its shells and food for its workers.

The lack of subsistence for the employees became a problem that both the City of Columbus and the Confederate supply systems could not resolve.

Major F. C. Humphreys, Chief Ordnance Officer for the Columbus Arsenal resorted to exchanging powder for bacon, yet he never resolved this problem as a scarcity of food for his workers haunted him throughout the war.

While Columbus had the potential to produce more than all the other Confederate arsenals, the combined and consistently acute shortages of food and lead held them back. (*Daily Sun* April 8, 1863; November 1864; Standard, 42)

Across the river in Girard, Alabama, the government employed workers in a Rope Factory, two blacksmith shops, the Mobile and Girard Rail Road roundhouse facility, and a Confederate Nail Factory.

CHAPTER 4
COLUMBUS, CONFEDERATE BASTION

In addition to the numerous manufacturing firms already mentioned, the most daunting of them all was the Columbus Iron Works, which contracted with the Confederate Ordnance Department to produce cannons. This foundry cast brass barrels, mortar tubes, and wrought iron rifled cannons.

In the spring of 1862, Major James H. Warner was assigned to Columbus to lease the large foundry for the sole use of the Confederate States Navy. Warner was to organize the facility and supervise the construction of Confederate gunboats. Safely located deep within Confederate lines, "Port Columbus" could easily build and deploy gunboats down the Chattahoochee River and into the Gulf.

As the commander of the Columbus Naval Iron Works, Major Warner supervised the construction of cannons, boilers, propellers, shafts, and all necessary shipbuilding tasks to build gunboats. By 1864, Columbus was building not only boilers for its own production, but also provided them for other naval projects. (OR Series II, Vol. 2, pp. 750-3; *Enquirer* November 6, 1864)

The naval facility at Columbus constructed several ships including two ironclads, the *CSS Jackson* and *CSS Muscogee*, a wooden gunboat, the *CSS Chattahoochee*, as well as the *CSS Shamrock* and the *CSS Viper*. These ships were constructed by 150 citizen/ sailors of Columbus, who were constantly plagued with a shortage of raw materials. While the facility continued its efforts, it resorted to accepting local implements made of metal to continue production.

The ironclad gunboat, CSS Jackson at Port Columbus

    The ladies of Columbus led the community in the collection of church bells, lamp stands, irons, and various household ornaments, which were all gathered and smelted for the war effort. (Enquirer, March 23, 1862; Sun, June 5, 1862; Standard 43) At least 80 brass cannons were built at the foundry and several are still in existence.

    The National Park Service owns several cannons built at Columbus and they grace the battlefield parks at Gettysburg and Shiloh. Another one dubbed the "Ladies Defender" has survived the ravages of time as well.

    The community poured its efforts into the Confederate cause and all male workers in the Confederate facilities also served as the local home guard. For example, the Stanford family of Girard, Alabama had a father and two of his sons employed at the Naval Works.

Thomas Stanford worked as the senior blacksmith and his sons Billy and John Henry worked as boilermakers. The youngest Stanford, Billy made $3.00 a day, the older and stronger boy, John made $5.00 per day, and their father, being a skilled employee made $10.00 for a day's pay. They also drilled twice a week as sailors in Company C, of the CSN Naval Iron Works Battalion commanded by their co-worker and commanding officer, Captain Christopher "C.C." McGehee. (Stanford letter; Navy Grey; CSU roster)

These citizen soldiers and sailors were proud of their service to both production and defense. As previously stated, Mayor Francis Wilkins, himself a wounded veteran, set the example for other casualties of the war. Despite their inability to serve in the line due to disabilities, many men joined the ranks of the local home guard. Like Billy Stanford serving in the CSN and working at the Naval Works, young men who were not yet old enough for the front served alongside the elderly and disabled.

As the war progressed, the Union war machine strengthened its grip on the Confederacy. By July of 1863, the Union had achieved great victories at Gettysburg and Vicksburg. The casualties of war affected every home, and Columbus was soon filled with hospitals, invalids, and indigent soldiers. This combined with inflation, poverty, and continuous military losses created a nervous anxiety amongst the people of the South and in Columbus home defense was questioned.

Appealing to the personal manhood and military traditions, which the community had always revered, the men that were able to shoulder a musket answered the call. As early as July of 1863, the local press lauded the response of her men to muster for service. The Columbus Enquirer reported:

We are happy to say that the turnout yesterday afternoon, in response to the call for a general

> *organization of our citizens for home defense, was fully up to the anticipation of the most sanguine. Seven or eight companies convened at the Court-house... the companies present formed a regiment electing Captain Francis Wilkins, Colonel... The proceedings and demonstrations of the day evinced a united determination of all our people to rally and organize for the defense of their homes.* (CE July 21, 1863)

Initially, Major Humphreys of the Ordnance Department and Major Dillard of the Quartermaster Department participated in the organization of the home guard, but they were far too busy with their own responsibilities to accept the post of Commandant for the Columbus garrison. The city's first official commandant was Captain C. B. Mims, another of the many well qualified, but disabled war veterans. (CE July 28, 1863)

After the serious Confederate losses at Gettysburg and Vicksburg in July of 1863, the Confederacy was forced to restructure its forces. Many of the Confederacy's key officers were disabled, discouraged or dead. By September of 1863, another young Captain, J. S. Smith assumed the duties of garrison commander as both the Confederacy and Georgia Militia restructured its forces.

Another soldier that received a new assignment was Howell Cobb. Transferred from a field command to a post in his home state of Georgia, Major General Cobb was assigned to command the newly formed Military District of Georgia where he would manage affairs for both the Confederacy and the State of Georgia. (OR Series I, Vol. 39, Part 2, p. 882; Vol. 45, Part 2, p. 644)

CSA Major General Howell Cobb

This important position was established and assigned to Cobb for several important reasons. First, Cobb was not a professional military man. While he was administratively proficient, knowledgeable, and a brilliant politician, Cobb was a weak tactician and poorly suited for arduous service in the field. Moreover, Cobb was in poor condition, overweight, and field service was rapidly wearing down his aging body.

Several of his peers, Robert Toombs and Gustavus W. Smith had resigned their regular Confederate commissions and joined the State Militia of Georgia.

Toombs served as the Militia's Chief of Staff, while Major General Smith commanded the Georgia Militia.

Thus, in Georgia, Cobb would lead men with whom he could readily command. Secondly, political affairs between Georgia and the Confederate administration had reached epic proportions as Georgia authorities clashed with the Davis administration.

As Union soldiers began threatening to invade the State of Georgia, Governor Joseph Brown resisted the call for more Georgians to serve in the Confederacy, and he worked diligently to retain them at home.

Thus, Cobb the consummate politician was forced into a situation where he served as the defacto mediator between Governor Brown and President Davis. While Cobb and Brown did not get along well, at least they were both Georgians, and few people got along with Governor Brown anyway.

The frustrations and heartaches inherent in a lengthy war affected not only the Confederate hierarchy, but the citizen/ soldiers in small communities as well. With a young captain assigned as the Post Commandant of Columbus, the men who outranked him discounted his authority.

An incident that highlights this situation concerned the simple matter of guard duty and a reluctance of some old veterans to perform the tasks as ordered.

The local paper reported that at least one veteran refused the task, stating that even if he was given "a hundred dollars in gold, he would leave the Confederacy before he would perform guard duty." The same article urged soldiers to acknowledge the authority of Captain

Smith and "as good soldiers, and patriotic citizens, they should go ahead, do their duty, and cease their everlasting grumblings." (CE November 10, 1863)

Another officer that had returned to Georgia after several years of service was Colonel Martin J. Crawford, former commander of the 3d Georgia Cavalry. Upon his return to Columbus he was assigned as a voluntary aide-de-camp for General Howell Cobb. Crawford kept Cobb informed concerning the various military affairs in Columbus, and he assisted in the management of administrative duties such as inspections of local troops and advisory duties.

Both men were deeply concerned about the ability of the Columbus garrison to defend itself from an attack. Cobb wrote to General P.G.T. Beauregard concerning Columbus, and he in turn responded by sending an expert to evaluate the situation.

In December of 1863, General Beauregard sent General Jeremy F. Gilmer to inspect both the organization of home defense forces and to make recommendations for earthworks with which to defend key locations. Columbus was valuable to the Confederacy and Gilmer returned a detailed synopsis of the situation at Columbus.

His report concerning troop availability in Columbus is listed in Table 1:

Table 1: Columbus Defense Troops (December 1863)

COMMAND	STRENGTH	REMARKS
Major Howard's Battalion	350	Guarding Government shops and works.
Lt. Colonel Thompson's Battalion	350	Local exempts. Drilling once or twice a week.
Naval Iron Works Battalion	150	Employees in naval workshops. Drilling once or twice a week.
Vigilance Fire Company	50	Local exempts. Drilling once or twice a week.
Regiment State Troops, Colonel Salisbury	400	Not yet called together.
Captain Chapman's Company	50	Local exempts. Drilling once or twice a week.
Ordnance Battalion, Major Baldwin	150	Employees in ordnance shops. Drilled.
Captain Latham's light battery	100	Local exempts. Armed with four 6-pounders, fully

	manned and depending on the horses of the town.
Total:	**1,500**

General Gilmer also reported that at least 25 men of Howard's Battalion could be mounted to serve as scouts. He also reported that Captain Smith was a fine young officer, but Gilmer recommended replacing him with an older officer of greater rank to better control the large garrison. (OR Series I, Vol. 28, part II, p. 554)

Beauregard responded immediately to Gilmer's recommendations and he sent Colonel J. W. Robertson to command the post and a young engineer, Captain Theodore Moreno to construct the requisite earthworks. (Ibid, p. 581)

Colonel Robertson quickly established a strong chain of command and improved the local defense force while Captain Moreno went to work on the fortifications.

Captain Theodore Moreno was an engineer by trade and an experienced Confederate Engineer. He was a graduate of the University of Virginia and he hailed from Pensacola, Florida.

Moreno had several brothers serving in the Confederacy and his sister Angela was the wife of Stephen Mallory, Secretary of the Confederate States Navy. (Saltmarsh letter, Mandrell 344-50)

CSA Captain Theodore Moreno

Captain Moreno had previously supervised the construction of Chattahoochee River defenses near Apalachicola, Florida and his abilities as an engineer were highly regarded. Preceding Moreno's arrival in Columbus was the following message from Confederate authorities:

> *The attention of the general commanding has been called to the importance of erecting works for the defense of Columbus, Ga. He has, therefore, directed Captain Moreno, of the Engineers; to repair there and construct such works for its efficient defense, as far as may be practicable and necessary. He desires that you give Captain Moreno every facility and all the aid in your power to enable him to prosecute the work before him.* (Series I, Vol. 14, Part 1, p. 681- 682; Series I, Vol. 28, part 2, p. 279, Series I, Vol. 52, Part 2, p. 375)

Moreno's family accompanied him to his new post, and they were warmly received in Columbus. Every necessity to assist Moreno in his duties was supplied as ordered and the year of 1864 was filled with the construction of numerous earthworks surrounding the Columbus area. (Mandrell, p. 344)

Moreno supervised the construction of a line of breastworks that encircled Columbus to the west and north, but the greatest emphasis was placed on fortifications that ringed the heights west of the Chattahoochee above the town of Girard. These forts were completed in late April of 1864, and primarily built using slave labor, which was generously provided by local plantations. (CE April 28, 1864)

The defense works west of Girard were built to guard the three wagon roads and a single rail line leading to the Columbus/ Girard area from points west in Alabama. Except for a foot bridge at Clapp's Factory, three miles north of town, these routes terminated in the Mill Creek (also referred to as Holland Creek) Valley of Girard where they merged into two wagon bridges and a railroad bridge that crossed the Chattahoochee River. These bridges provided direct passage into Georgia and would be of great strategic importance in the months ahead.

The forts that guarded these roads consisted of eight independent forts placed strategically on high ground overlooking approaches to the city. The forts were designed of primarily three main types (square, pentagonal, and curved) and connected by a series of breastworks. These forts were constructed with continuous bankettes, deep ditches and at least one bombproof.

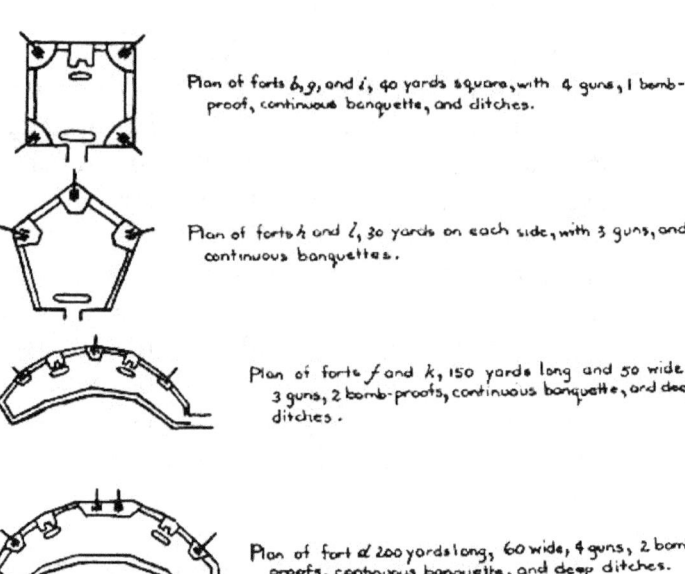

Confederate Forts constructed to defend Columbus/Girard

The square forts were 40 yards long on each side capable of mounting 4 guns each, and the pentagonal forts were 30 yards long per side with 3 gun positions. The curved forts were 150 yards long by 50 yards wide with 3 gun positions, except for the fort on Ingersoll Hill, which was 200 yards long and 60 yards wide with 4 gun positions.

The defense network was expertly constructed, but if properly manned, they would have required every man, woman and child in the community.

Still, it was prudent to prepare for a "worst-case scenario," and as if in a premonition of events yet to unfold, Gilmer told Moreno that "the town and public property may at no distant day be exposed to raids of mounted troops." (OR Series 1, Vol. 28, part 2, page 554) Yet, in the early months of 1864, the locals considered the construction of such extensive earthworks as being unnecessary, but prudent.

Only time and the threat of Union aggression could validate their true value. Moreno was commended for his work by the traditional Confederate reward, harder work and more responsibility as the Chief Engineer, Military District of Georgia, and Third District of South Carolina. (Series 2, Vol. 7, p. 518-19)

During May of 1864, the Columbus garrison was again reorganized. Colonel Robertson was reassigned and replaced by Major George C. Dawson. Throughout the war, local troops were organized and disbanded, transferred, and ultimately, they created a confusing dilemma for Confederate authorities.

Thus, Howard's Battalion was disbanded due to its questionable composition, which upon inspection contained a predominant ratio of boys under the age of seventeen. Initially formed by Major Thacker B. Howard and registered with Confederate authorities as the 27th Georgia Battalion, the unit appeared on the rolls as a valid fighting force; however, upon examination, it contained only locals who were either too old or too young (non-conscripts) to fight. (CE May 20, 1864; June 10, 1864)

Other units existed in Columbus, and one of these was known as Jacques Battalion of Reserves, which consisted of local factory workers. This organization was originally a part of Lt. Colonel D. B. Thompson's Battalion, also confusingly known as the 19th Battalion Georgia Infantry, State Guards, the

1st City Battalion, the City Guard Battalion, and the 1st Muscogee Battalion. Nonetheless, "Jacques Battalion of Reserves" formed by Samuel R. Jacques, factory supervisor, was organized better than most factory worker / combat units, and apparently, they were eager to join their brothers-in-arms as they were all volunteers. At least the men drilled on a regular basis and they contained six full companies of men derived from the factories of Columbus.

These hard-working laborer/ soldiers were organized by companies according to the factory in which they worked. At least one other unit, known as the 5th Regiment Georgia Infantry Reserves also drilled in Columbus, and it too was comprised of local factory workers.

The ranks of the local Arsenal Battalion thinned out in 1864 as the arsenal's capacity to manufacture ammunition dropped off due to a shortage of raw materials. Still, the unit maintained a force of 32 men that drilled and remained prepared for defense work if called to duty.

Lieutenant Charles M. Kinsel who worked for the Haiman brothers as an engraver led this unit. The Columbus Fire Guards served as local firemen and they too also drilled as a military unit, under their commander and Fire Chief, Captain W. C. Gray. One other local unit that reformed in the summer of 1864 was a small detachment of cavalrymen, known as the "Muscogee Cavalry."

Comprised of 57 experienced veterans, these men provided the local area with a scouting capability, and Captain John S. Pemberton commanded them. (Muscogee County Court Records- Local Troop rosters, Author's note: John S. Pemberton survived the war, he became a pharmacist and spent many years focusing his work on developing pain relief medications. In 1886, he invented Coca-Cola).

CSA Captain John S. Pemberton

Another Georgia unit that played a key role in events that unfolded at Columbus concerns the unit affectionately or contemptuously (depending on one's perspective) referred to as "Joe Brown's Army," or "Pets".

In the latter years of the war, Governor Joseph Brown passively resisted the Confederate Conscription Act of February 17, 1864, by "exempting" the bulk of his own state employees. He also retained at least two full regiments of State Guards or Reserves, who were properly known as the "Georgia State Line."

These two regiments designated the 1st and 2nd regiments of the Georgia State Line were comprised of 10 regulation-size companies and they participated in several engagements throughout the war. Originally organized as railroad guards, the troops enlisted to defend their home state of Georgia, but on several occasions assisted regular Confederate forces beyond the borders of the state.

The Georgia State Line was amongst the best troops held in the Confederate Reserve, but again, they answered first to the Governor of Georgia, then to Confederate authorities. (Bragg x, xi, 128-161)

While Georgia contributed more than her fair share of men to the Confederacy, the latter years of the war were filled with clashes between state and Confederate authorities. The Confederacy literally ran out of manpower and the states clambered for men to defend their homes as Union forces threatened them.

Governor Joe Brown was openly critical of the Confederate government after Sherman tore into his state. Bitter words were exchanged between General Howell Cobb, Governor Brown, and Confederate authorities concerning the management of scarce resources; however, all of them were guilty of making irrational decisions.

General Cobb as commander of the Military District of Georgia was placed in a "no win" situation as he tried to appease both Confederate authorities and meet the demands of his own state. In June of 1864 Cobb tried to mediate the matter of conscription between his governor and Confederate authorities.

Governor Brown accused General Cobb of spreading "malicious slander" about him and his efforts to exempt many of Georgia's men from compulsory military duty. Cobb then relayed Brown's blistering letter to the Confederate Secretary of War James Seddon stating, "This communication and the author are alike unworthy of further notice." (Series 4, Vol. 3, p. 55)

Across the river in Alabama, Governor Thomas H. Watts denounced the Conscription Act of 1864 and the steady erosion of his own reserve forces. Writing to Confederate authorities, Watts stated:

> *I think it the most egregious folly... These volunteer companies of which I speak have been organized in their respective counties and they are ready to respond whenever a call for the reserves is made. In the meantime, except when called out to drill, which is once a week, they are permitted to remain at home making something to eat for their families and for our soldiers... I cannot permit the troops organized for State defense and ready to obey my calls to be all taken out of the control of the State. The laws of Alabama must be executed and I must have some troops at my command to execute them.* (Series 4, Vol. 3, p. 463)

General John Preston, Superintendent of the Bureau of Conscription pleaded with President Davis and Secretary Seddon with, "If the Governors persist in raising and holding armies composed largely of classes brought into the service of the Confederacy by the act of 17th February 1864, then the anticipated effect of that act is neutralized." (Series 4, Vol. 3, p. 463)

Ultimately, neither state nor Confederate authorities resolved the matter and local conscription officers simply took matters into their own hands. Depending upon the officer's loyalties or motives, he either passively resisted his orders or overzealously pursued them. The result was men or boys were stripped from badly needed home guards and shipped to front-line duty, and the poor soldier was either too young, too old, or incapable of being an effective soldier. Likewise, some communities failed to provide conscripts that were available.

Thus, while the Confederacy unraveled, the Union war machine, which included Wilson's Cavalry Corps, pressed harder and faster towards the annihilation of them all, and with equal bias.

While the Georgia State Line would contribute significantly to the defense of Columbus, the burden of preparation fell squarely on the men of the local community. As the war progressed, the duties of the Columbus garrison grew as the local community was constantly expanding, but not with more men capable of fighting.

Several additional hospitals were opened in Columbus after the destruction of Atlanta, and they remained busy until the war's end. While the government shops absorbed some of the recuperating soldiers in its work force over the next few months, these predominantly "invalid" men were incapable of contributing to front line duties.

Tragically, Provost Marshall duties were a constant task as broken men sought solace in "the bottle," although the sale of liquor was forbidden.

The local papers frequently noted these societal woes and the "disorderly and lawless conduct of the hospital soldiers in Columbus" became a big problem. (CE July 30, 1864) To quell this problem and to improve good order and discipline in the post at Columbus, a new commandant was assigned on September 9, 1864. (CE 9/9/64)

Colonel Leon Von Zinken was the son of a Prussian general and he had served in the Prussian army prior to immigrating to the United States. He settled in Louisiana and when the Civil War began, he was commissioned a Major in the 20th Louisiana Infantry Regiment.

Colonel Von Zinken successfully commanded his regiment in the Battle of Perryville in 1862, and he was promoted to the rank of Lieutenant Colonel. In 1863, he was promoted to the rank of Colonel and he led his regiment in battle at Chickamauga, Missionary Ridge and throughout the Atlanta campaign until he suffered a debilitating wound at the Battle of Ezra Church on July 28, 1864.

Suffering from a shattered arm, he was transferred from the Army of Tennessee to garrison duties in Georgia. Von Zinken was a professional soldier and he was described as "energetic, enterprising, and a spirited officer." (CE September 11, 1864)

The new post commander immediately established the most rigid adherence to discipline that Columbus had ever experienced. Von Zinken noted that "emissaries and spies of the enemy" were at work in the area and he established an identification system that required all citizens to carry a pass at all times. (CE September 12, 1864) He also cracked down on the "hospital soldiers" who plagued the city with their rowdy, off-time leisure activities.

Several officers assisted Colonel Von Zinken with the myriad duties of running such a large military post. Like their commander, Captain Isidore Guilliet and Major Samuel L. Bishop, were also from Louisiana, and had they had served under his command for three years. All these men bore the scars of battle, and they ran an efficient and disciplined post. Captain Guillet served as Von Zinken's Adjutant, and Major Bishop performed the duties of Provost Marshall.

Major Bishop's duties included the management of the Provost Guard to actively police the streets; check passes, and maintain security throughout the city.

Captain Thomas E. Blanchard, a local officer convalescing at home due to wounds received at the Battle of Resaca, assisted

Major Bishop with the provost duties. In 1861, at Montgomery, Blanchard had fired the first salute honoring President Jefferson Davis, and ever since that day, he had fought to preserve the Confederacy.

Blanchard was a proud young man, and like so many of his fellow soldiers, although he was crippled by war, he performed what duties he could. Described as a highly capable and energetic soldier, Blanchard had risen through the ranks from Private to Captain. He led men of the 37th Georgia Infantry into battle, and he had served on the staff of General G.W.C. Lee. While Bishop and Blanchard performed the duties of Provost Marshall, Colonel Von Zinken's "right hand man" was his adjutant, Captain Isidore Guillet.

With the numerous facilities, continuous operations, personnel, requisitions, communications, logistics, etc., all conducted during a civilian community, Captain Guillet was in great demand, every day, and 24 hours per day. Guillet was of French descent and he hailed from a proud Louisiana family that contributed four brothers to the Confederacy.

After three years of warfare, Isidore was the sole survivor. His most recent loss had occurred at the battle of Murfreesboro, where he witnessed the death of his eldest brother, Major Charles Guillet. Armed with capable and devoted assistants, Colonel Von Zinken focused his efforts on organizing his post. (George Greene Collection- Von Zinken papers)

On September 16, 1864, Colonel Von Zinken called upon the community to help him establish more breastworks around the city and better connect the established fortifications. In this task, the citizens of Columbus eagerly assisted Von Zinken and they praised his work. Von Zinken instilled order in both the post and community, and he established his headquarters on Broad Street between 12th and 13th Streets.

There was little doubt who was in charge after Von Zinken placed a row of 12-pound howitzers in front of his headquarters in downtown Columbus. Thus, atop the building formerly occupied by a popular civilian shoe store, Von Zinken raised his garrison flag, a huge 8 by 13 foot, Second National Confederate flag. While the citizens of Columbus initially welcomed a heightened sense of security, a certain level of anxiety must have been felt as well.

A local writer recorded that, when Von Zinken was asked why he emplaced the cannons downtown, the commander's reply in broken English was, "Vell, if tem dam Yankees come here I make vun 'ell of a tam fuss!" While Colonel Von Zinken was a capable officer, the militaristic antics of the red-haired Prussian officer must have been somewhat amusing to the locals. After all, the enemy had never really threatened Columbus, but the locals failed to grasp that Colonel Von Zinken was serious in his work. (Telfair, 133; Worsley, 294)

In Columbus, the year 1865 began with hopes that the insanity of war would soon end. Instead, it began with a tragic incident that tested Colonel Von Zinken's authority and his relations with the people of Columbus. Despite the measures Von Zinken took to improve security conditions in Columbus, he was consistently haunted by the threat of subversion, espionage, and enemy incursion. Such are the burdens of command, and Von Zinken's were huge.

During the long cold months of winter, soldiers reach their zenith of misery, and are not at their best. Such was the case when a popular young soldier, Private John Lindsay of the 17th Georgia, Benning's Brigade, Lee's Army of Northern Virginia returned home to Columbus on wounded furlough. As he entered town, the local guard demanded to see his papers, and the veteran soldier contemptuously ignored the repeated orders from the zealous guard. Colonel Von Zinken was informed of the infraction and he ordered Lindsay brought to him dead or alive.

The next day Lindsay passed by the same guard, and the scene was repeated, but this time the guard shot and killed the young man. A mob of citizens set out to lynch Von Zinken, and if not for the arrival of Lindsay's heartbroken, but compassionate father, he may have been hung. Colonel Von Zinken was charged as an accessory to murder, but a military court-martial led by General Howell Cobb acquitted him. (GA Archives, Pension Records, Widow of John Lindsay; Worsley, 293)

Von Zinken's authority and his strict rules had been challenged. While the incident was a tragic affair, the general populace learned a hard lesson in the serious nature of discipline and military regulations amidst martial law.

Colonel Von Zinken's pass and identification policy was a prudent initiative amidst the myriad threats of war, and as the epic events of 1865 unfolded, the threats and perceptions thereof brought chaos and stress to a new level. Confederate authorities had long suspected that pro-union sympathizers lived and worked in the Columbus area.

One such man was Mr. Randolph Mott, a wealthy elder of Columbus. Mott was a well-respected business tycoon and generally regarded as a pillar of the community. Even though his son, Captain John R. Mott served in General Henry Benning's brigade and fought for the Confederacy; throughout the war, the elder Mott proudly boasted after the fall of Columbus that he never left the Union.

Mott also showed the good sense of selling most of the production of his Palace Mills production of wheat and corn to the Confederate government. While the citizens of Columbus generally considered Mott to be harmless, and they respected his opinions, there were other pro-unionists that were not trusted, and for good cause. To root out traitors in the Columbus are, a "Vigilance Committee" was formed and pro-Unionists faced the scrutiny of their peers.

A secret pro-Union organization operated throughout the war against the Confederacy, and it was very active in and around the Army of Tennessee, and the states of Alabama and Georgia. Known as the "Peace Society," its members lived and worked in Southern communities and they actively plotted acts of subversion and espionage directed against the Confederacy. President Davis reported to the Confederate Congress that:

> *Secret leagues and associations are being formed. In certain localities men of no mean position do not hesitate to avow their disloyalty and hostility to our cause... In districts overrun by the enemy or liable to their encroachments, citizens of well-known disloyalty are holding frequent communication with them, and furnishing valuable information to our injury, even to the frustration of important military movements.* (OR Series 4, Vol. 3, p.67)

While most of the alleged work of the "Peace Society" occurred in Alabama, there were suspected members of this organization that operated in Georgia. In 1864, a secret special investigation was conducted by Confederate investigators and the alleged members included "amongst them Hon. James Johnston [sic] and Dr. Tuggle, of Columbus, Ga., and George Reese, of West Point." (OR Series 4, Vol. 3, 393-96; Series 1, Vol. 39, Part 2, p. 585-90)

While these men were never formally charged, nor did the Confederate authorities ever prove that they were members of the "Peace Society," some historians suspect that Johnson may have been an active participant. Johnson was a long-time political nemesis of Henry Benning and Howell Cobb. While he had served as a local attorney and a US Congressman from Georgia in 1851 through 1853, Johnson was generally disliked and ostracized from Columbus social circles. (George Greene Collection)

In recent years, several researchers have highlighted Johnson's alleged ties with the "Peace Society," and several post-war narratives written by Union officers bear witness that a local "citizen" assisted them with information concerning Confederate positions and facilities in and around Columbus. (CLE Bill Winn; Scott, Upton Report)

Johnson also owned a house in Russell County, Alabama, from which he moved and promptly sold just prior to the Union attack. (CDS March-April 1865) After the war, James Johnson, Dr. Thomas Tuggle, and Randolph Mott admitted that they were close friends throughout the war and they often discussed the war and political matters, yet there has never been any evidence found that links them to subversion or espionage directed against the Confederacy. (National Archives, Record Group 217, Microfiche # M1658, Fiche # 1, 2, & 3; Claim Number 15615)

Moreover, Johnson, Tuggle, and Mott were never positively linked to the "Peace Society" and its activities to subvert the Confederacy in Columbus; however, once the war was won, Johnson was suddenly appointed as the Provisional Governor of Georgia. During a post-war speech, with General James H. Wilson by his side, Johnson publicly ridiculed his fellow Georgians for their participation in the former Confederacy as a "stupendous folly." (Dameron, 305)

With the alleged conspiracies and myriad challenges of command, Colonel Von Zinken was heavily burdened in his duties at Columbus; however, as the epic events of 1865 unfolded, he would test his mettle as never before.

In January of 1865, General P.G.T. Beauregard and his staff studied the likely course of events that the Union would employ in the western theater. With the Confederacy being threatened on all fronts, their defensive strategy would be crucial in the months ahead. General Beauregard provided President Davis with an intelligence estimate that proved to be extremely accurate. On January 24, 1865, he wrote:

> *Present appearances indicate following as early plan of campaign on part of enemy: Thomas, from Middle Tennessee, will move via Eastport, Tuscumbia, or Decatur into North Alabama, on Selma or Montgomery. Canby will move via Pascagoula, Mobile, and South Alabama to form a junction with Thomas. They will then probably march on Columbus and Macon. General Taylor will have, to oppose these two armies, only Stewart's corps, about 5,000 men, Mobile garrison about same, Forrest and other troops about 10,000 more.* (OR Series 1, Vol. 49, Part I, p. 929)

In February, General Taylor telegraphed General Cobb with a cautionary message that stated, "In view of present contingencies it is important to fortify and garrison Columbus." Cobb expressed concern that he lacked adequate troops to perform the mission and requested additional information concerning the perceived threat. (OR Series 1, Vol. 49, Part I, p. 1011-12)

General Taylor also sent a message to Major General Samuel Jones, then commander of the Military District of South Georgia and Florida to, "Please send all surplus arms in your district to Major-General Cobb. They are absolutely required to arm the garrison at Columbus, Ga., which must be defended on account of the large Government factories." (OR Series 1, Vol. 49, Part I, p. 1041)

But in 1865, neither General Jones nor any other Confederate unit had any "surplus arms" to send anywhere, and before any more attention could be placed on Columbus, reports were received that Union forces were marching into Alabama. Taylor called on the cavalry of General Nathan B. Forrest to determine the enemy's intent, and he began preparations to safeguard Confederate property in Alabama.

Protecting the railways was of paramount importance and he had his staff ascertain "how long it will take to move the engines from the Montgomery and Mobile Road to Columbus, Ga., and have preparations secretly made to do it if it becomes necessary." (OR Series 1, Vol. 49, Part I p. 1023)

As Wilson's raiders struck at the heart of Alabama, Von Zinken found himself literally in the middle, between General Cobb in Macon, and his Confederate brethren fighting the enemy in Alabama.

Telegraphic messages, couriers, trains, troops, and supplies all made their way through Columbus as the enemy pressed ever closer to Von Zinken's post. Yet, for Von Zinken, rumors served as the most vexing and materially damaging item that zipped through his post.

For a military commander charged with the defense of a city, quelling a nervous public was a full-time job. He had good relations with the local press who reflected generally positive news despite serious threats in neighboring Alabama.

One of the first in a series of articles that encouraged the citizens of Columbus to think positive thoughts despite alleged rumors of Union victories in Alabama appeared in the *Columbus Daily Sun* on March 31, 1865.

The alleged bad news of Wilson cutting a swath across Alabama was ridiculed as "mere Northern telegraphic speculation" and besides, "Forrest, 'The Wizard of the Saddle' is on the war path, and the indomitable Buford is wakeful and watchful." (CDS March 31, 1865)

Buford was on the run. Forrest was wounded, and his command had suffered a series of brutal and rapid defeats at the hands of Wilson's Cavalry Corps. Forrest's cavalrymen had fought the Union raiders at Montevallo, Randolph, Ebenezer Church, and Selma, Alabama.

Each time they clashed with Wilson's cavalry, the Confederates were badly beaten, and General Wilson's Union raiders continued riding right across Alabama, and Columbus, Georgia was next on the list.

On April 2, 1865, Colonel Von Zinken repeated a plea for permission to allow Negroes to help defend Columbus. For many Southerners, the concept of arming black men was contradictory to their ideals, but not for practical men such as Colonel Von Zinken. He telegraphed Confederate authorities with, "Many Negroes offered daily to volunteer. Could raise a brigade in a short time. Have telegraphed twice on the subject. Please answer." (OR Series 1, Vol. 49, Part 2, p. 1193.)

While, the Confederacy had passed legislation and begun to arm blacks in Virginia, the initiative was authorized too late in the outlying departments, thus Von Zinken never received the reply he so desperately sought. General Taylor was also eager to arm Negro volunteers, but his reply was, "No orders from the proper authorities at Richmond have as yet reached him on the subject of the late legislation with regard to the employment of Negroes as soldiers, but this would not prove an obstacle with the commanding general in the acceptance of this proposition." (OR Series 1, Vol. 49, Part 2, p. 1199)

Nonetheless, increased manpower was an imminent need. While the fortifications that lined Columbus and Girard were extensive, and there was no way Colonel Von Zinken could fill them with men.

Meanwhile, the local press again attempted to reassure their readers with:

> *Alarmists- There are in every community those who grow tremulous and excited at the remotest prospect of danger... They seem to have forgotten that Sherman with his entire army was much nearer Columbus six months ago; that the whole State was at the mercy of the enemy. Things are somewhat different now. The city of Columbus is defended today by a much more efficient force than at any previous time, while the prospect of attack is diminished... A sufficient force will be sent here in time to defend it against any Yankee raid... Everything is in readiness. Let us keep cool.* (CDS April 4, 1865)

Colonel Von Zinken coordinated what defense measures he could and it was clear that Wilson's Union corps would approach Columbus from the west. While he knew that manning the fortifications built by Major Moreno guarding the main routes into the area was warranted; with limited manpower, he would have to focus his defense on his enemy's key targets, the bridges leading into town.

Still, Girard, Alabama would have to provide a barrier and the first line of defense between the Union raiders and the bridges.

Colonel Von Zinken's counterpart in Girard was Colonel Lyman W. Martin. Martin's unit was referred to as the Russell County "Home Guard," which consisted of two local defense companies, "Perry's Reserves" and the "Ist Class Militia." These two companies comprised a total of 120 men, of which 40% were 16 years old. (AA RH-46: SG 24912)

The composition of these units mirrored those in Columbus, and while some of the men were factory workers, the bulk of the units consisted of those who were too young, too old, or invalids. When the war began, most of Russell County's men eagerly volunteered for the Confederacy and through attrition, most of them would never return.

In 1865, Colonel Martin was a tough leader, but he bore the scars of battle, and had returned home due to disabilities. Still, he attempted to contribute his remaining days and energies to the defense of his home state. Colonel Von Zinken requested that he mobilize his companies for an impending battle, but the man was too ill, thus command devolved to his deputy commander, Colonel John Brannon.

Colonel Brannon was the principal of a local Alabama school, and when he was summoned to report to Von Zinken, he dismissed his students with, "Children, gather up your book and slates and go quietly to your homes. It is rumored that a Yankee raid is coming in this direction and I am ordered to get the reserves ready to meet it." (CLE April 11, 1865; Russell County History, P. C-45; CLE February 5, 1965; Walker, 191 & 205)

One of Brannon's students was Charles Martin who had just turned fourteen. He recalled that after Colonel Brannon dismissed the students, "An elder sister, a younger brother, and I proceeded to our home, arriving there about twelve o'clock. We found the citizens of the village all in a hustle of excitement, packing knapsacks with clothing and filling haversacks with rations for the members of the families who were expecting to be ordered to the front at any minute." (Russell County History, P. C-45)

After he sent the children home, Brannon formed his "men" known as the Russell County "Home Guards" and preparations were begun for the impending battle. The local Alabamians worked alongside their Georgian brothers-in-arms, digging trenches, constructing abatis, and placing "slashings" of trees, which were cut and emplaced in front of the earthworks.

Among the Russell County soldiers," working in the trenches were three other members of the Martin clan. Lieutenant William N. Martin and his sons, William W. and Charles, the fourteen-year-old student of Colonel Brannon, were assigned to the trench in front (west wall) of the fort on Red Hill.

In this location, these men were destined to see a lot of action in the hours ahead. William W. Martin was the only veteran in the family, having served in Lee's Army of Northern Virginia for several years and had only returned home after receiving his fifth wound, which was in the process of healing. (Russell County History, P. C-45, F-154)

Still, the shortage of men was acute and Von Zinken appealed to everyone he could for assistance. Again, the local press assisted Von Zinken by calling for everyone in town to "delay no longer… organize and protect your homes." Moreover, the press also encouraged their fellow citizens with, "never let it be said that Columbus fell without a struggle." (CDS April 8, 1865)

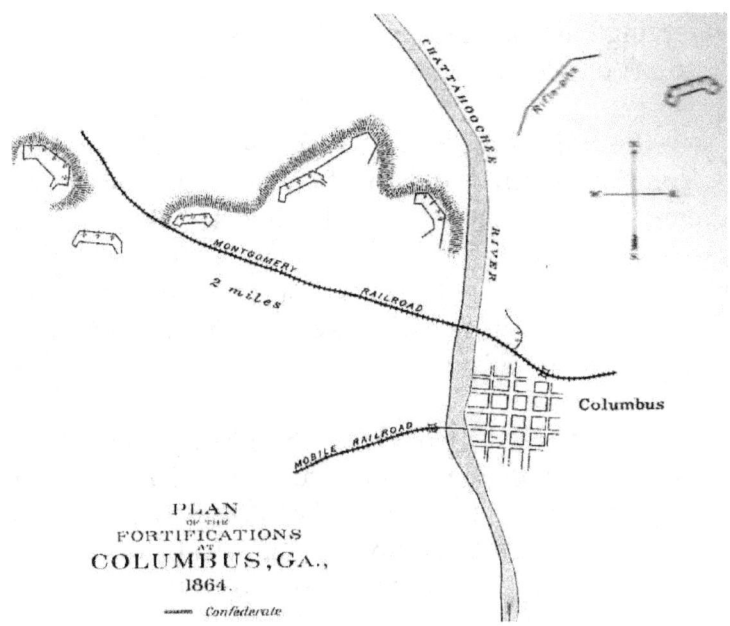
Confederate fortifications at Columbus/ Girard

To assist in the defense of Columbus and Girard, Colonel James C. Cole, commander of the post at Opelika, Alabama deployed with his battalion (company strength) where they joined Colonel Brannon's troops. Colonel Cole was a disabled Confederate officer and he was originally from Tennessee. His background was like Colonel Von Zinken in that he was disabled while serving in the Army of Tennessee, and he became a post commander.

Cole began his service as a Major in the 21st Tennessee Infantry Regiment, but he received rapid promotions and he commanded the 3d and 5th Confederate Infantry Regiment, which was a consolidated unit consisting of various companies from Alabama, Tennessee, Florida, Kentucky, and Arkansas.

During the Chattanooga Campaign, Cole was so severely wounded that he was not expected to live. Despite his wounds, Cole survived, and although he was disabled, he continued serving the Confederacy. (OR Series 1, Vol. 31, part II, p. 660, 762; Vol. 52, Part II, p. 793)

From his Post Headquarters on Broad Street, Colonel Von Zinken directed daily affairs and he initiated an emergency alarm system that would signal impending danger. Upon the firing of six cannon shots, all military organizations would rally at designated locations "fully armed and equipped, with twenty-four hours rations, ready to take the field."

The press also noted that "Some of our people are excited," but there was no factual news from Montgomery, only "extravagant rumors." The citizens were encouraged to remain "calm and rational," and "ready to aid in the defense of the city." (CDS April 10, 1865)

The local populace was also informed that the alarm meant for them to seek shelter, and they were reminded that upon the issuance of the signal, all liquor establishments were to close immediately, as the sale of spirits would be suspended. (CDS April 15, 1865)

General Cobb realized the gravity of the impending threat to Georgia and he contacted President Davis seeking guidance. Their communications reflect the sounds of a dying administration. Cobb warned, "The movements in Alabama put in immediate danger arsenals and public stores at Columbus… I should know whether other forces will be sent here." (Series 1, Vol. 49, Part 2, p. 1208)

By this time, President Davis had fled Richmond and was hiding in Danville, Virginia and being pursued by General Grant. From that location, he replied to Cobb that Governor Watts in Alabama was also seeking guidance, and if Cobb could, please contact Governor Watts and "as far as practicable to aid in the defense of Alabama. Communicate with him."

President Davis then ceased his efforts to help them stating, "May God bless your efforts and give you success."

From his headquarters in Macon, Georgia, General Cobb had no way of knowing that the Confederacy was being rolled up on all fronts, but in a final plea for help from the Confederacy, Cobb contacted General Beauregard who promptly replied that he should "call on the Governor of Georgia for all assistance practicable." (Series 1, Vol. 49, Part 2, p. 1212)

Realizing that there were no Confederate forces available to assist him, General Cobb went to Columbus to confer with Colonel Von Zinken. On April 11, Cobb telegraphed his Georgia troops at Jackson's Station, just below Macon, and he ordered the 1st and 2nd Regiments of the Georgia State Line to move immediately via rail for Columbus. Cobb also received several detachments of General Wofford's reserves that deployed from Macon and Augusta.

Cobb notified Governor Brown that the Georgia Militia needed to be called out in force to protect Georgia from an imminent invasion. Governor Brown was reluctant to call out the militia, en masse, and on this important decision, he chose to wait. (Bragg, 107)

By this time, however, General Wilson had shattered Forrest's Cavalry Corps, destroyed Selma, accepted the surrender of Montgomery, and he was bearing down hard on Columbus.

On April 12, 1865, the local press provided a brief article stating that Montgomery had been evacuated, but it ended with the word, "RUMOR." (CDS April 12, 1865)

Late the next day, refugees and soldiers from Montgomery, Alabama arrived in Columbus, and the mystery was solved. The alleged crisis was real. The invading "Yankee horde" had smashed Selma, Montgomery had surrendered, and the "extravagant rumor" was factual truth. Among the men that arrived from Montgomery were Governor Thomas Watts, and two Confederate generals, Abraham Buford and Daniel Adams. Watts planned to move down river and reestablish the Alabama State government in Eufaula.

General Buford, whose cavalrymen were harassing the advancing Union column, would remain and assist in the defense of Columbus. Artillery and Infantrymen commanded by General Adams arrived by train, and they would provide great assistance in the days ahead. While these troops were experienced veterans, the fighting in Alabama had decimated their units. Cobb's men of the Georgia State Line also arrived via rail from Lovejoy Station, and they too were assigned positions within the fortifications in Girard or as reserves in Columbus.

Even though three Confederate generals were present, the senior leader, General Howell Cobb gave the overall command of troops to Colonel Von Zinken. Whether this move was a statement of confidence for Von Zinken or an act of plausible deniability in case of failure, Cobb publicly announced his decision in the afternoon edition of the Columbus Daily Sun. His statement reads:

All doubt is at an end. The enemy is advancing rapidly upon your city. Every arrangement that could be made for your defense has been and is being made. A stern resistance saves your city, and arrests the progress of the enemy through your country. I appeal to the citizens of Columbus and the neighboring counties of Georgia and Alabama, to respond at once to the call now made upon them to defend themselves, their homes and their country.

Let every man capable of bearing arms, report promptly for duty, so that he may strike an honest blow for all that is dear to freemen. Prompt and energetic action is alone necessary to secure success. Bring with you whatever arms you have, and those who have none will be supplied.... Colonel Von Zinken will take command of all troops in and around Columbus. (CDS April 15, 1865)

The paper also listed a statement from Colonel Von Zinken, which echoed the appeal for citizens to arm themselves, and an order suspending all businesses from operation. On the same date that Columbus braced for attack, Governor Brown ceased bickering with Confederate authorities and he announced a call-up of the entire Georgia Militia.

Colonel Von Zinken's public notification of the impending Union attack on Columbus

In an urgent message to General Smith to prepare the militia for action, his order was issued too late to help in the defense of Georgia. Brown's message states:

The movements of the enemy in Central Alabama indicate an intention on their part to make an early movement upon Columbus and other points in Georgia. To enable us to meet this successfully, it will require the united efforts of all who are able to bear arms, whether they belong to the State or Confederate service. You, are, therefore, hereby directed to order out the militia of the State, subject to your command, to rendezvous at Columbus, as fast as possible.

All who are subject to your command under your former orders from these headquarters are embraced in this call, and all subject to militia duty under fifty years of age who fall to respond will be turned over to Confederate service. I regret exceedingly to have to require them to leave their corps at this important period, but the movement of the enemy leaves no other alternative. (OR Series 1, Vol. 52, Part 2 Chapter LXIV, "Supplements", p. 813)

The governor's message carried the intent to arm nearly every man in Georgia, but again, the order was issued too late as Wilson's raiders were within several miles of the state line.

That afternoon, April 15, 1865, all work ceased in the area factories, and workers were dismissed to join their respective commands in the field. On this date, the logbook maintained at the Columbus Navy yard states, "The enemy reported 12 miles from here- the troops ordered out to the trenches." (Turner, 233)

Unbeknownst to these part-time sailors, they would see action in the impending battle. With three companies, they were assigned to defend both their naval facility and the Columbus side of the Dillingham Street (lower) bridge. Major Warner had four cannons emplaced and manned along the banks of the Chattahoochee River in the northern end of the Naval works.

He also provided Captain "C.C." McGehee's company with four 6-pound brass cannons, which they emplaced behind a parapet on the Georgia side of their assigned bridge. McGehee had served several years in the army as an artilleryman before joining the Navy, thus he was well prepared for this mission. Both sets of cannons were oriented towards the Alabama side of the river and prepared to fire at anyone threatening the bridge or Naval facility.

Planks were removed from the bridge in a large section on the Alabama side so that it could not be crossed. Spaces between the remaining planks were stuffed with cotton, and the entire bridge was saturated with turpentine and kerosene. McGehee's orders were to destroy the bridge if threatened by a Union assault.

Major Warner, commander of the facility and troops ordered four naval vessels, the *Mariana, Young, Viper* and *Chattahoochee* to move a safe distance down river (12-15 miles) from Columbus; however, the uncompleted gunboat, *C.S.S. Jackson* would remain in port.

While the Navy was responsible for the lower bridge and the southern end of Columbus, everything else including the railroad bridge, the 14th Street (upper) bridge, and all fortifications west of Columbus had to be defended and those bridges prepared for destruction as well. Thus, in town, the two wagon bridges and the single railroad bridge were all defended by cannons and prepared for emergency destruction, if needed.

Private Washington Crumpton, a local member of the militia relates his role in the preparations with, "After an all-night job, cutting ropes on cotton bales so they'd burn easily, we were sent to Girard on the Georgia line. All day we were piling barrels of rosin and fat lightwood under the bridges." (Crumpton, p. 100)

Planking from the footbridge at Clapp's Factory, three miles north of town was removed to prevent a Union crossing at that point. Colonel Von Zinken's primary tools for defense would be his cannons. Within the Columbus foundry and arsenal, Von Zinken had at least 74 cannons in town, and a virtually unlimited supply of shells.

His challenge was to get them placed and manned wisely. Von Zinken strategically emplaced 14 cannons on the Columbus side of the river and 24 across the river in Alabama. Like the four cannons guarding the Dillingham Street bridge, Von Zinken emplaced five cannons at the railroad bridge behind a protective berm, and two cannons were placed in the road at the head of the 14th Street bridge. Again, these cannons were located on the Columbus side of the river and oriented to the west.

Directly across the 14th Street bridge, and two blocks west was a large square-shaped earthen fort atop a prominent landmark, known as Red Hill. From this position, the fort's four 12- pound howitzers could sweep the entire length of Mill Creek Valley to the west, Girard and the Dillingham Street Bridge to the south, northward along the approaches of Summerville Road, and eastward it covered the 14th Street Bridge and downtown Columbus.

This important fortification was the base of Von Zinken's defensive strategy, and since the arrival of General Adams' forces, he now had one of the Confederacy's best artillery officers to command it, Major James Fleming Waddell. Originally from Russell County, Waddell was a prominent citizen and his wife; Julia was the daughter of Dr. Edwin deGraffenried, a founding father of Columbus. Most of Waddell's 140 men were also from Russell County, and they would staunchly defend both their assigned positions and their homes.

Waddell's Alabama Artillery Battery was originally formed in 1862, from selected members of the 6th Alabama Regiment. The battery participated in the Battle of Shiloh and the Kentucky Campaign. It also fought at the Siege of Vicksburg, where it was captured on July 4, 1863. The unit was paroled and exchanged 2 months later, but after returning home, they reformed their unit in Columbus, Georgia. Armed with 10-

lb. Parrott guns, the unit consisted of two companies, which were commanded by Captain Winslow D. Emery, Company "A", and Captain Richard H. Bellamy, Company "B".

Waddell's newly formed unit was then assigned to the Army of Tennessee, where it was combined with another battery, the 10th Missouri. Commanded by Captain Overton Barrett, the battery was comprised of four 12-lb. Howitzers, but during the Atlanta Campaign, it was bolstered by a section of Captain Herman Sengstak's Alabama Battery. This unit was disbanded after the loss at Vicksburg, Mississippi, thus its surviving members were incorporated into the Army of Tennessee in the 10th Missouri and Waddell's Battalion.

The 10th Missouri Battery had seen considerable action at Corinth, Perryville, Tullahoma, Chickamauga, and Chattanooga. Major Waddell was given command of these batteries and his unit was re-designated as the 20th Alabama, Light Artillery Battalion. Waddell's 20th Artillery Battalion then served with distinction at the battle of Murfreesboro and throughout the fight for Atlanta.

After Hood's Army of Tennessee was reorganized in 1864, Waddell's two batteries commanded by Captains Emery and Bellamy were assigned to the Department of Alabama and served with General Daniel W. Adams.

During the winter of 1864, the 10th Missouri was detached from Waddell's unit and posted independently at Macon and Columbus, Georgia. Rejoined and assigned with a new mission, Waddell's old 20th Artillery Battalion was ready for work. These artillerymen were confident, competent, and used to working together. Thus, Colonel Von Zinken placed Major Waddell in charge of all the artillery.

Waddell strengthened the fortification on Red Hill with at least six additional cannons, which were placed in a row facing west along the north exterior wall of the fort. In this corner, the row of guns guarded the Alabama side of the 14th Street Bridge. These guns could also sweep westward, deeply into the Mill Creek Valley, and if needed, they could traverse uphill to the northwest covering the approach along the Summerville Road. The guns were also protected by a trench line to their front (west) that ran from their left and adjacent to the fort on Red Hill, and extending right (northward) and parallel to the Summerville Road.

Captain Richard Bellamy and Major James Waddell

This road runs uphill to the village of Summerville (3 miles beyond Girard), and a trench line extending 1.7 miles ran parallel to this road. Within this trench line, Waddell emplaced a 2-gun lunette and rifle pits in strategic locations to cover the Summerville Road. Extending to the top of Ingersoll Hill, this trench line intersected with a perpendicular trench line that contained 3 more forts.

These forts created an exterior line that guarded against any incursion from the north along Summerville Road. Consisting of two curved forts on either side of the road and terminating in its westernmost extension with a large fort like the one on Red Hill, these forts and their interconnecting east-west trench line, were 1 mile in length.

These upper works (forts and lunettes) on Ingersoll Hill contained cannons although the two forts west of Summerville Road had no guns. Strategically placed and well-constructed, Colonel Von Zinken manned these upper forts with another Alabama artillery unit that had arrived with General Adams from Montgomery. Commanded by Captain Nathaniel (Nate) H. Clanton, this unit of veteran artillerymen was strengthened with additional men and equipment.

While Clanton was originally from Macon, Georgia, his unit was formed in Montgomery County, Alabama in March of 1863. Captain Clanton's battery was assigned to a brigade commanded by his brother, General James H. Clanton. The unit fought throughout Alabama until Clanton's brigade was nearly destroyed in battle near Pollard, Alabama (Bluff Springs) on March 25, 1865.

The remaining sections of his battery were then attached to General Daniel Adams, whose units retreated eastward before Wilson's cavalrymen. The battery consisted of two 12-lb. Howitzers, three 6-lb. Smoothbores and one Parrott. (Sisafikis; Dyer; Grant) Unbeknownst to these men who were emplaced upon seemingly lonely heights far from the important bridges below, they would soon face a direct assault from Wilson's Cavalry Corps. (Sifakis; Dyer)

Within the trenches, Colonel Von Zinken placed as many men as he could muster. Complementing his rather abundant supply of artillerymen, cannons, and shells, he had a viable force of infantrymen as well.

From Georgia, he had the two regiments of the Georgia State Line and various local reserve battalions. From Alabama, he had the Russell County "Home Guard," and Cole's Battalion, commanded by Colonel Brannon, and the amalgamated Infantry force of General Daniel Adams.

This composite patchwork of men claimed membership in units that no longer existed, and likely, they were its sole survivors. Adams' unit contained soldiers that represented Alabama, Mississippi, Louisiana, Tennessee, Arkansas, and Texas. Nonetheless, they were Confederates with a will to fight, and that was good enough for General Adams, and Von Zinken certainly appreciated them as well.

Conservative estimates and fragmentary reports indicate that by April 16, 1865, Adams' foot soldiers did not exceed 1,000 men. (Columbus hospital reports, various OR)

Rounding out the composite forces assembling to defend Columbus were General Buford's cavalrymen. Abraham Buford had led cavalrymen since he graduated from West Point in 1841. He was a veteran of the Mexican War where he served as a dragoon. While he was originally from Kentucky and the brother of General John W. Buford of the Union cavalry, Abraham was an old warrior, and a staunch Confederate.

As a division commander under General Nathan Bedford Forrest, Buford had commanded several brigades of Kentucky and Alabama cavalrymen; however, as the war wound down, few men remained from his original command.

In April of 1865, the bulk of his unit consisted primarily of the 7th Alabama Cavalry Regiment. This unit was a zealous group of warriors and they were all expert horsemen. Organized in July of 1863 from counties in and around Montgomery and Pollard, the 7th Alabama Cavalry regiment was initially assigned to General James H. Clanton's brigade.

The unit served throughout Alabama until the fall of 1864 when the regiment was reassigned to Rucker's brigade of Forrest's Cavalry Corps. The regiment served with distinction under Forrest, and saw heavy fighting that depleted the unit to just 64 men. The unit was then reorganized and after several months of recruiting during the winter of 1864-1865, it raised its strength to 300 men.

They were then transferred to the command of General Abraham Buford, of Forrest's Cavalry Corps. In 1865, as Wilson's raiders cut across Alabama, the 7th Alabama regiment continuously harassed Wilson's Union raiders all the way from Selma, Alabama to Columbus, Georgia.

The 7th was joined by remnants of the 6th and 8th Alabama Cavalry regiments, who were the remaining men of Armistead and Clanton's brigades. Several other cavalrymen from the 10th Texas & 2nd Alabama were also included in this composite cavalry force.

While their former units no longer existed, the Confederate cavalrymen eagerly served under General Buford, and they diligently continued the fight.

These various units comprised a force of 300 men, but at least 75% of them were men of the 7th Alabama Cavalry regiment. When General Buford arrived in Columbus, Captain John Pemberton's force of 57 cavalrymen attached themselves to Buford's unit, and rode east down the Crawford Road to help harass the Union column.

As Buford's cavalrymen fell back into Columbus from their hit and run strikes against the advancing Union column, he divided them into three detachments. One was placed within the defense lines on the Alabama side of the river, while the other two were strategically placed along the east banks of the Chattahoochee, above (north) and (south) below Columbus to guard against any Union river crossing and a flanking maneuver into Columbus. (Hoole, p. 15)

In addition to the forces called up for the defense of Columbus, one other organization joined the Confederate defenders just in time for the impending battle. On the morning of April 16, the 26th Battalion of the 66th Georgia regiment commanded by Major John W. Nisbet arrived in Columbus.

While Nisbet's unit was on a special assignment and simply passing through Columbus, General Cobb gave the young major a personal request "to help the citizens in their defense of the city." Major Nisbet was the nephew of Georgia Supreme Court Justice Eugenius Nisbet, and his brother was a Confederate officer as well. He had entered military service as a private infantryman, and despite ill health due to exposure in the field, he excelled as a soldier.

According to Nisbet's commander, "His men adored him, not only for his lenient temper, but for his big heart. As a private, his self-will was unruffled, as there was no danger of being cashiered. But as an officer, queerly enough, he continued to know no will but his own." (Nisbet, 256)

Colonel Von Zinken was pleased to welcome the 26th Battalion to his command, and according to Major Nisbet, "My battalion was posted in Girard, Alabama, across the Chattahoochee. We were in the center, citizens being on my right and left, their flanks resting on the river, above and below the bridge." (Nisbet, 256)

Thus, as the impending battle loomed before him, Colonel Von Zinken's defense force consisted of an estimated 3,250 men. The following table represents the units, their strength, and assigned positions during the Battle of Columbus.

Table 2: Confederate Troops (Order of Battle)

COMMAND	STRENGTH	ASSIGNED POSITION
INFANTRY		
Jacques Battalion (Major Samuel R. Jacques)	150	Trench line, left & right wings, (Summerville Road)
Russell County "Home Guard" Reserves (Colonel John Brannon)	120	In fort on Red Hill, trench line adjacent to fort and at the 14th Street Bridge (Alabama side of river).
Cole's Battalion "Opelika Volunteers" (Colonel James C. Cole)	120	Reserve force from Opelika, Alabama. In trenches with Russell County troops (Alabama side of river).
Naval Iron Works Battalion (Major J. H. Warner)	150	Naval port facility (Iron Works) at riverbank and Dillingham Street (lower)

Unit	Strength	Location
Columbus Arsenal Battalion (Lieutenant Charles M. Kinsel)	32	Bridge (Columbus side of river). Trench line on Summerville Road (Girard, Alabama)
Columbus "Fire Guards" (Captain W.C. Gray)	46	Columbus, Georgia (Reserves/ Firemen)
1st Regiment Georgia State Line (Colonel Edward Galt)	500	Columbus, Georgia (Reserve Force)
2nd Regiment Georgia State Line (Colonel Richard Storey)	500	Throughout earthworks & forts (Alabama side of river).
Composite Alabama Infantry (General Daniel Adams)	1,000	Throughout earthworks & forts (Alabama side of river).
Naval Iron Works Battalion	50	Center of trench line along Summerville Road (Alabama side of river).

CAVALRY		
Composite Alabama Cavalry (General Abraham Buford)	300	3 detachments (1- filled gaps in the main trench line, 2 detachments provided flank security)
"Pemberton's Cavalry" (Captain John S. Pemberton)	57	Near fort on Red Hill (Girard, Alabama)
ARTILLERY		
Waddell's Alabama Artillery (Major James Waddell)	100	Fort on Red Hill, (Girard, Alabama)
10th Missouri Battery (Captain Overton Barrett)	50	North wall of trench line at Red Hill (Girard, Alabama).
Clanton's Alabama Battery (Captain Nathan Clanton)	50	Fort and lunette on Ingersoll Hill (Alabama side of river).
Columbus Light Battery (Captain Latham)	25	Railroad bridge and 14th Street bridge

		(Columbus, Georgia)
Total:	**3,250**	

Upton's 4th Division Cavalry

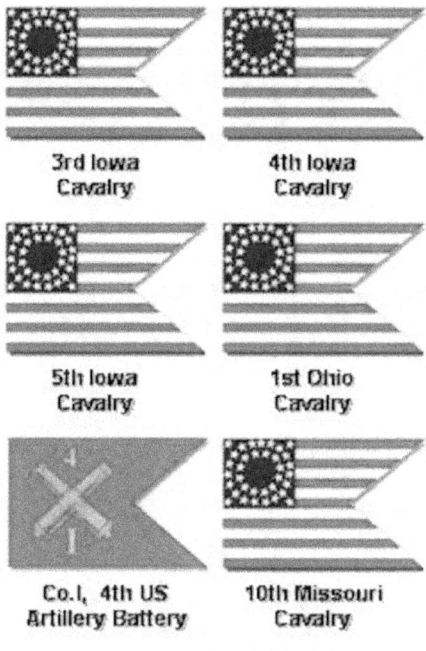

3rd Iowa Cavalry

4th Iowa Cavalry

5th Iowa Cavalry

1st Ohio Cavalry

Co.I, 4th US Artillery Battery

10th Missouri Cavalry

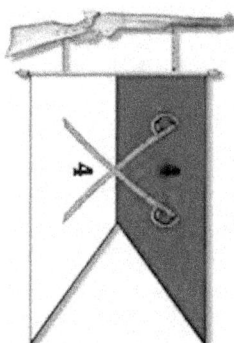

Wilson's Cavalry Corps

Union

Von Zinken's Defense Force

Buford's Cavalry

Georgia State Line (2 Regiments)

Austin's Battalion (14th Louisiana)

7th Alabama Cavalry

Cole's Alabama Local Defense Force

Russell County, Alabama Reserves

10th Missouri Artillery Battery

Waddell's Artillery Battery

Co. A, 27th Battalion Georgia Volunteers

Adams' Alabama State Reserves

Confederate

CHAPTER 5
THE BATTLE

Colonel Von Zinken's command had increased considerably in just a few short days, and on the morning of battle, his men strengthened the trench line along the Summerville Road. The breastworks were dug deeply enough to accommodate the tallest men and parapets were bolstered to provide maximum protection.

Yet, one of the men who would ultimately man that trench, Private Washington Crumpton of the Georgia militia ominously noted that, "A straight ditch, dug by the militia, up and down the hill, would afford no protection, once the line was broken at the top of the hill. So, we prepared to run a foot race for the bridges, if the line should break." (Crumpton, p.100)

Being unable to man the outer fortifications west of town along the Crawford, Salem, and Opelika Roads, would ultimately haunt the Confederate defenders as well. With just skirmishers between Girard and the approaching Union column, the enemy would be directly upon them in short order.

Yet, the inner line of works surrounding the 14th Street Bridge formed a "tête de pont" (fortified bridgehead) that would be hard to penetrate, especially protected by an abundance of artillery. Von Zinken also had his men strengthen the inner line by cutting and placing abatis and "slashings" to a depth of 100 yards in front of the earthworks.

At the Naval Iron Works, the daily logbook entry states, "Easter Sunday- weather pleasant- The excitement in town intence [sic] - Began early this morning loading stores on *Chattahoochee & Jackson*." (Turner, p. 233)

While the soldiers and sailors made last minute improvements on their positions and prepared for the enemy's arrival, the local citizens went to church and prayed for God's deliverance and mercy. These stalwart Christians displayed a defiant confidence as one lady recorded that before battle on that Sunday morning, "the Episcopalians and Roman Catholics went to early Communion to pray for victory." (Worsley, 294)

Meanwhile, on Crawford Road, several miles west of Girard, Buford's cavalrymen completed their final ambush on the advancing Union column. Riding back to Girard with the news of the enemy's imminent arrival, General Buford received their report and placed them in position for the fight.

Von Zinken's provost assistant, Captain Thomas Blanchard was then dispatched into Columbus, and "like Paul Revere, he rode a horse through the town to notify the inhabitants." (Worsley, 294)

While confidence in the citizenry had not yet waned, it was suddenly challenged as Ms. Louis Gunby Jones, a young lady of fourteen on that fateful day wrote, "About ten o'clock Sunday morning the outposts across the river rushed into Columbus and roused the already expectant populace into a furor of excitement." (Jones, CSU paper)

After the abrupt church service, Mrs. James (Laura) Comer rushed home to her plantation where she recorded in her dairy, "When I left Columbus intense excitement prevailed. The federals were said to have captured Montgomery & were approaching Columbus… What a terrible thing <u>war</u> is!" (Comer Diary, CSU notes)

The citizens of Girard heeded the warning and evacuated the area, as their homes would be in the direct line of fire. Mrs. Thomas Stanton hastily loaded a wagon with a few personal possessions. With her husband and sons serving across the river in the Naval Battalion, she and her three little girls hurriedly evacuated towards Columbus. Before reaching the bridge, however, the horse suddenly stopped, and died.

Mrs. Stanton and her little girls rushed back to their home, and clamored under their house. Huddled together, the terrified girls waited for what fate and the Yankees would bring to Girard. (Stanton letter)

By noon, Wilson's Cavalry Corps had penetrated deeply into Russell County, Alabama and they were closing in on their target. Pressing forward with cautious determination Upton's advanced scouts of Alexander's brigade had exchanged fire with Confederate cavalrymen throughout the morning, but the fire suddenly ended as they neared the crest of a hill.

As the Union scouts crossed the hill, they could see the enemy fleeing into a valley below them. They could also see the town of Girard, Alabama, the Chattahoochee River, and their objective, Columbus, Georgia. General Upton moved forward and through his field glasses, he surveyed the vicinity surrounding the twin cities of Girard and Columbus.

While the main target was of great interest to him, Upton focused on his immediate needs first, the bridges across the Chattahoochee.

General Wilson had provided Upton with a challenging mission, and to achieve it, he needed to take just one bridge, but it had to be intact. While Upton could clearly see the three bridges before him, he also saw a formidable array of cannons, trenches, forts, and his enemy within them. With awkward surrealistic stares the combatants gazed upon one another and one of Wilson's troopers recorded the scene in his own words with:

From our position overlooking the whole valley for miles up and down the river we could see everything plain and distinct. The marching of the troops into the forts and breastworks - the gunners loading their Heavy Guns and their Officers, Generals & Staff riding on full gallop along their lines giving their orders, and all their other preparations.

We could see the people - men, women and children in Columbus rushing to and fro and even see them pointing their spy glasses towards us - it was a curious and wonderful sight to us - and grand beyond description was the view of the river valley and city in its martial array - But just now we did not appreciate all this grandeur. Our thoughts were busy thinking of how we were ever going to get over them 3 or 4 hundred yards that separated us from our friends in gray - alive, as soon as we came in view and they saw us on those hills.

The bells were set ringing - Whistles from factories shrieked their loudest, and warning salutes were fired to summon the people to arms and the defense of their homes against the brutal invaders. It seemed that all pandemonium had broken loose. (Conzett, 76)

Ominously, the Confederates withheld their fire, while setting their sights and measuring the range to their foe. Wasting no time, General Upton had Colonel Beroth B. Eggleston, commanding the 1st Ohio Cavalry Regiment form an assault force to charge the lower bridge.

The Dillingham Street Bridge lay directly before them, and Eggleston selected six companies and quickly organized them for the attack. The time was now 2:00 P.M., on April 16, 1865, Easter Sunday. The mounted column of Ohio cavalrymen formed by fours and prepared for battle. It was a beautiful spring day, and the sun bathed the blue-clad warriors in warm, soft light as a gentle breeze lightly caressed the wispy manes and tails of their trusted steeds.

The peaceful serenity was abruptly ended, as General Upton challenged his men with, "Can you give us the bridge across the Chattahoochee?" The proud men of Ohio saluted their commander and answered him with "We will try." Captain Yeoman of the 1st Ohio recorded, "Straight away down the hill lay the bridge across the Chattahoochee. It was the prize we sought; could we take it?" (Yeoman, p. 223)

Colonel Eggleston and his staff took their positions at the head of the column. He then ordered his men to draw their sabers, and the bugler sounded the order to "Charge." Captain Yeoman recorded that first the column rode out at "the trot, then the gallop, then the wild charge, and away we went, straight down for Columbus!" (Yeoman, p. 223)

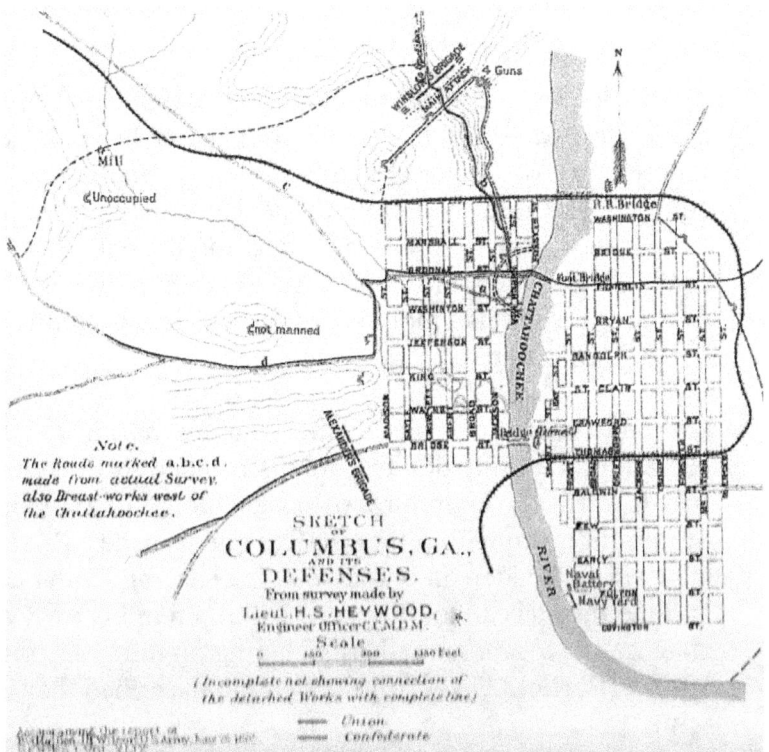
Confederate Defenses of Columbus/ Girard

Observing the 1st Ohio Cavalry Regiment charging downward into the tiny hamlet of Girard, Alabama, was their commander, General Emory Upton. From a high wooded ridge west of town, the general and his staff witnessed the beginning of their final battle against the Confederacy.

Eager to treat Columbus like they had done Selma, Upton's division watched from their reserve positions along the ridges west of Girard. With great anticipation, the Union troopers watched as their fellow soldiers dashed forward on a grand cavalry charge.

General Upton peered forward as his Ohioans wound their way downward into Mill Creek Valley, and he gleefully remarked, "Columbus is ours without firing a shot!" (Yeoman,

p. 223)

Suddenly, the silence cloaking the defenders of Columbus and Girard was abruptly shattered as Major James Waddell ordered the howitzers on Red Hill to open fire. Upton's adjutant, Ebenezer N. Gilpin of the Third Iowa, recorded, "We were standing on a little knoll, watching the enemy across the bridge, and as they did not fire began to think the place was evacuated when in a moment, every gun in Columbus opened on us." (Gilpin, p. 652)

Unleashing solid shot at their maximum range and elevation, the Confederate cannons belched forth their violent fury towards the Union troops posted along the ridgeline west of Girard. From within the Confederate trenches, Private Charles K. Henderson witnessed the charging Union cavalry and the furious cannon fire, and described in his own words, he stated, "I saw cannon balls ricocheting on the sides of Girardean Hills." (Barfield, 752) Adjutant Gilpin described their effect with:

> *The shot came fast and furious. Two of our headquarters horses were killed. One shell struck our chief bugler's horse, tearing him all to pieces. Then grape and canister, more than I ever want to hear again. More horses were killed, but fortunately none of us.* (Gilpin, 652)

As the cavalrymen swept into the streets of Girard, the Confederate artillery did not follow, lest they blast into their own homes. However, across the river on Bridge Street, the Confederate battery led by Captain "C.C." McGehee waited for the optimum moment to unleash a barrage of grape and canister into the approaching Union cavalry.

At the lead of the charging Union column, Captain Yeoman noted that the cannons guarding the Dillingham Street Bridge had not opened fire. Realizing that they must be loaded with grape and canister, he pointed it out to his commander, Colonel Eggleston. Faced with the reality of charging directly across the bridge and being shredded by Confederate fire, Colonel Eggleston signaled the column and:

> *Ordered a left oblique on the impulse of the moment, and we rode in on the pavement in shelter of the houses, while orders were at once given to dismount and fight on foot. While these preparations were going on, it was suggested to throw a few sharpshooters at this end of the bridge to cover the guns and shoot down the gunners while we were forming in front of it for the charge.* (Yeoman, p. 223)

According to the Ohioans, one of their sharpshooters cut down a Confederate artilleryman that pulled the lanyard as he was falling, thus igniting the bridge into a wall of flames. According to Confederate reports, the bridge was destroyed "with conspicuous gallantry" by Captain "C. C." McGehee of the Naval Battalion. "Several men were killed and several wounded on each side in this first brush with the enemy." (Worsley, p. 295)

In whatever manner and by whom it was destroyed is irrelevant as the bridge was immediately engulfed with flames and rendered impassable. The action happened so quickly "that the men standing at this end of the bridge caught the breath of the fire in their very faces. This change in the method of attack saved the whole battalion; for whatever number of the battalion had attempted to cross the bridge, no man would have ever gotten out of it alive."

At the lead of the charge and catching the full blast of the inferno was Captain James Kirkendall, commander of company D, displaying his "usual coolness and gallantry". (Yeoman, p. 224) Had they made it in their charge for the bridge, Kirkendall and his men would have been roasted alive, and they were extremely agitated with the Confederates for their choice of tactics.

One of Kirkendall's men, Sergeant Aden Harper "rode out of the little town of Girard and across the bridge over the little creek which separated the earthworks from the town of Girard, and rode almost up to the works, and then rode back again. It was a gallant and an Inspiring thing to do." (Yeoman, p. 225)

While Harper's impetuous charge achieved nothing more than praise from his comrades in Company D, the Confederates could clearly see that the Union troops were a determined foe.

From their position at the intersection of Bridge and Abert Streets in Girard, the Union cavalrymen turned southwest and rode uphill to the high ground along Sandfort Road, where General Alexander's entire brigade rallied and established their headquarters.

In this location, Alexander's 2nd Brigade would occupy the heights overlooking the lower bridge, while Captain George B. Rodney's artillery battery went into action.

As the Union artillery exchanged fire with the Confederate artillerists, Waddell's battery on Red Hill, fired extensively, and during the exchange, an exploding shell destroyed one of his gun crews. The horrible scene was witnessed and described by the Martin clan, who were hugging the earth on Red Hill. The explosion destroyed the cannon and splattered its crew throughout the fort.

The resulting carnage of blood and gore unnerved and depressed Lieutenant William N. Martin and his youngest son, Charles, who reflected that "my brother who was used to such sights laughed at them, and told them that this was nothing to what they would see in a short time." (Russell County History,

p. C-46)

The veteran was correct and shortly after this explosion another one validated his remark. Again, as the Martin clan huddled close to the earth, George Hooper Waddell rode up to the Martins' position and yelled, "Men, this position must be held at all hazards." Then recognizing Lieutenant Martin, Waddell declared, "Martin, I know that you will hold it!"

George Waddell was a Russell County judge and the brother of Major Waddell. On this day, however, instead of running the local court, he was assisting his brother maintain order in the heat of battle. After riding away from the Martin's position, their county judge was killed by an exploding cannon ball.

Meanwhile, the Union commanders watched the artillery duel with intense curiosity as they developed the precise locations of all the Confederate cannons. According to Adjutant Gilpin, "Glass in hand, General Upton stood like 'Patience on a monument,' scanning their position until satisfied it was impossible to attack successfully from that point, then ordered us to withdraw." (Gilpin, p. 652)

Satisfied that the objective could not be achieved through Mill Valley, and "impossible to attack the tête de pont from this direction," Upton ordered his 1st Brigade led by Colonel Edward Winslow into concealed positions beyond the western ridgeline. (Michie, p. 167) Upton also withdrew Rodney's Battery and after several hours of heavy artillery fire, the battlefield fell ominously silent.

General Upton immediately dispatched two detachments on special missions to the north of Girard. 200 men of the 10th Missouri Cavalry Regiment led by Captain Jeremiah F. Young was "ordered to ride with all speed" and seize the foot bridge north of Columbus at Clapp's Factory.

General Upton also sent companies A and F of the 5th Iowa Cavalry Regiment, led by Captain Edward Lewis to perform a reconnaissance of the Summerville Road and its approaches to the 14th Street Bridge.

Colonel Morris Young, 5th Iowa Cavalry

General Upton also divided his staff into scouting parties to perform a general survey of the northern approaches and get as close as possible to the enemy's defense works. Winslow's brigade moved northward behind the cover of the ridgeline crossing the Columbus and Western Railroad, Mill Creek, and halted in the woods near the roads leading to Salem and Opelika.

That afternoon while the Confederates waited for the next move of their adversary, the rest of Wilson's Cavalry Corps moved forward and joined Upton's division on the outskirts of Girard. General Wilson moved to the head of Winslow's column and as he approached General Upton, he observed the young general dashing off to the north with his reconnaissance parties.

General Wilson and the rest of the cavalry corps took a much-needed break while the Confederates continued pondering what the Yanks were up to. It was postulated that the Union artillery would mass along the ridgelines and Wilson would lay siege upon the Confederate bastion. (CE June 27, 1865; Martin, Part II, p. 179)

Their assumption was totally wrong and the silence was broken as Captain Young's Missourians tried to seize the bridge at Clapp's Factory. As planned, the Confederates had rendered the foot bridge impassable by removing its planking, while they focused their defense on the railroad and 14th Street bridges whose proximity to one another made them readily defensible. The Union cavalrymen now had to focus their efforts on these bridges to accomplish their goal.

Meanwhile, General Upton found a local "citizen" that provided information concerning the Confederate positions, their disposition, and strength. (Scott, 492) Again, it is unknown who this mysterious informant was and whether members of the Peace Society were involved remains an enigma.

During the reconnaissance patrols, several close calls resulted in the loss of two troopers that were taken prisoner by the Confederates. Major William Woods and Lieutenant Sloan Keck both lost men in their patrols, and one of them was Private Robert C. Wood of Company A, 4th Iowa Cavalry. Wood was taken across the river and held captive in the Mott house. (Scott, 491)

Satisfied that he could take the northern positions on Summerville Road, General Upton left Captain Lewis and the two companies of the 5th Iowa to serve as guides, and he rushed back to retrieve Winslow's brigade for the assault. Yet, Upton was rapidly losing daylight and complicating matters, he could not locate Colonel Winslow. Frantically searching for his men in the waning light and unfamiliar terrain, Upton rushed back and forth along the Opelika and Salem roads.

According to General Wilson, Upton's patrol approached his location and reported to his commander stating, "Everything is ready for the assault, but I cannot find Winslow and must delay the attack till he is in position." General Wilson pointed out the concealed location where Upton had placed Winslow's brigade, and Upton relied, "But it is now too late. It will be dark before I can get him into position and lead the division to the attack."

General Wilson listened intently to his young protégé and after determining that Upton had conceived a viable plan, Wilson made an important decision with epic consequences. He was confident in the capabilities of General Upton and his entire corps, thus he told Upton:

It is not too late to carry the plan into effect tonight; you will make all your arrangements to attack at 8:30. With flashing eyes, he exclaimed, "Do you mean it? It will be dark as midnight by that hour and that will be a night attack, indeed! "

Assuring him that it was just what I wanted and that it should be made, with Minty's division supporting, I instructed him to get everything ready to carry it in effect. With enthusiastic promptitude, he exclaimed, "By jingo I'll do it; and I'll sweep everything before me!" (Wilson, 259-60)

Wilson had the Union batteries fire occasional shots into the target area to distract the Confederates from their movement to the north. Meanwhile, "after all arrangements were completed, the men made coffee, got supper, and passed an hour in absolute rest, while Upton waited for the appointed minute with confidence and patience." (Wilson, 261)

According to Adjutant Gilpin, "From a hill, from which I could see every house in Columbus, every fort and earthwork, I watched the two armies maneuver until it was dark… A dim blue line of hills, as far as the eye can see, encircles the plain in which the city nestles. In the twilight, General Upton withdrew the First Brigade." (Gilpin, 653)

Before moving out, Colonel Winslow studied the sketch of the area that the "citizen" had prepared for the Union cavalrymen. The movement to the launch site was several miles away and the brigade swung around in a wide arc to the northeast until they connected with the Summerville Road.

The distances and key locations provided by the "citizen" proved to be "substantially accurate," and while it took about an hour to get into position, by 8:00 pm, the Union cavalrymen were poised for the assault. (Scott, p. 492)

According to Captain Charlie Hinricks of the 10th Missouri, "We layed [sic] low until after dark, then mounted without the bugle sound and we formed on a road leading as I supposed towards the town. We were ordered to be quiet, not to speak aloud and all noisy things such as tin baskets and pans were ordered to be thrown away. We halted and waited for the signal. It got dark as pitch, not a sound could be heard anywhere." (Hinricks, April 16)

The night was perfect for a stealthy attack. The sky was extremely dark and obscured by clouds, thus the only discernible object was the sandy road with its eerie white glow and the promise of danger in the distance. Yet, the darkness was their greatest ally, as a daylight attack would have already invoked deadly artillery fire. Creeping quietly along the Summerville Road, Winslow's brigade linked up with Captain

Lewis, who guided them to their positions, which had been predetermined by General Upton.

Captain Lewis and his two companies of the 5th Iowa would serve as skirmishers and lead the advancing force toward the enemy fortifications to their front. Behind them were six dismounted companies of the Third Iowa Cavalry, led by Colonel Noble.

His companies formed a line perpendicular to Summerville Road, with his left on the road and his right extending several hundred yards within the woods to the west. Lined up in column formation and four hundred yards behind the Third Cavalry was Colonel Frederick Benteen's 10th Missouri Cavalry.

Colonel Frederick Benteen,
Commander, 10th Missouri Cavalry Regiment

These troopers were all mounted, formed into a column of fours, and ready to charge as directed. 300 yards behind the Missourians were the mounted cavalrymen of the 4th Iowa Cavalry, commanded by Lieutenant Colonel John H. Peters. 200 yards in front of Lewis's skirmishers were the enemy, in two forts connected by a trench line and all perpendicular to the

Summerville Road.

Behind the forts, the terrain and the Summerville Road sloped downward for 1.7 miles to Red Hill and their prize, the 14th Street Bridge. But first, the Union cavalrymen would have to fight through the Confederate defenses built along the crest of the heights.

As the Union cavalrymen formed for the attack, looking towards their left front (southeast) was the most prominent terrain feature, Ingersoll Hill. This hill housed the largest fort on the heights and it was nearly 200 yards long. Within its earthen walls were four cannons and on either side of this fort there were lunettes mounting a total of three more.

Traversing to the right (west) from Ingersoll Hill and along the trench line to their front was another fort, and this one was rounded like its counterpart on Ingersoll Hill, but a little smaller with an overall length of 150 yards.

This fort did not contain any cannons, and the Union troopers would bear down heavily upon it. Continuing to the west, the terrain rapidly descended to a ravine, and sharply ascended to another prominent hilltop where there was a large, square- shaped fort like the bastion on Red Hill. This fort contained no artillery.

According to Colonel John Noble, the latter position "was a well-constructed fort (no guns being in position, however) flanked by well-constructed rifle-pits running nearly east." (OR Series 1, Vol. 49, Part 1, p. 493, Noble's Report)

Throughout the trench line and forts, Confederate troops were lined behind the parapets, and waiting for action. Between the Union and Confederate lines, were abatis and slashings that had been hastily prepared that very morning. The previously wooded area was now filled with stumps and felled trees pointing away from the forts, which provided a chaotic web of obstructions to anyone attempting to breech them.

General Upton planned for Alexander's 2nd brigade to simultaneously assault from his position southwest of Girard and attack the Confederate fort on Red Hill, but his primary focus was on the positions north of Girard, and he would personally command that endeavor.

General Upton had assembled 27 companies of cavalrymen for his night attack and moved them into location without being detected. This assault force consisted primarily of the 1st Brigade, 4th Division with two attached companies of the 5th Iowa Cavalry, from the 2nd Brigade. Thus, a total of 1,716 Union soldiers would be attacking a Confederate force nearly twice their size.

Table 3: Union Assault Force
(Order of Battle, April 16, 1865)

COMMAND	STRENGTH	REMARKS
3d Iowa Cavalry (6 companies- A, B, C, D, I, and K)	300	Dismounted and led by Colonel John W. Noble.
4th Iowa Cavalry (11 companies- entire regiment)	696	Mounted and led by Lieutenant Colonel John H. Peters.
5th Iowa Cavalry (2 companies- A and F)	200	Dismounted skirmishers led by Captain Edward Lewis.
10th Missouri Cavalry (8 companies- entire regiment)	520	Mounted and led by Lieutenant Colonel Frederick W. Benteen.
Total:	**1,716**	(OR Series 1, Vol. 49, Part 1, p. 483)

While the corps and division commanders, Generals Wilson and Upton were present on the field, Colonel Edward Winslow commanding the 1st Brigade and his regimental commanders, all had an awesome task ahead of them. Moreover, in a night attack, maneuver at the company level is critical to mission success, and throughout Wilson's Cavalry Corps, its company commanders were confident, competent, and experienced.

*Colonel Edward Winslow, Commander,
1st Brigade, Upton's Division*

The individual soldiers were also highly trained, proficient, and bolstered with high morale. While the Union assault force was nearly half the size of the force they were attacking, the Union plan was designed with but one objective, the seizure of the 14th Street Bridge. However, to reach their goal, they had to breech the Confederate line- punch a hole in the defenses, charge down to the river, and take the bridge.

General Upton's plan was to breech the exterior line with dismounted forces, and then, launch a "flying column" charging straight down the Summerville Road to the bridge. Upton assigned Captain Robert McGlasson with two companies of the 10th Missouri for this daring mission, and he placed them at the head of the mounted columns, so they would be in position to storm ahead on his command.

While there are conflicting reports as to the exact time that the battle commenced, it was between 8:00 and 9:00 pm when General Upton ordered Captain Lewis and his skirmishers forward, and his plan went into motion. According to Captain Harrington, the men were all ordered to yell as loudly as they could when the attack commenced to unnerve the enemy. (Harrington Diary, April 16, 1865)

According to Colonel Winslow, the opening scene of battle was fast, furious, and violent. *"The moment we were ready to attack, the enemy opened fire in front, with small-arms and on the left with shell, canister, and musketry, when the Third Iowa was directed to charge."* (OR Series 1, Vol. 49, Part 1, p. 481- Winslow's Report)

Sergeant Harrington of the 4th Iowa recorded that as the "charge commenced, and the woods rang again from the noise made by our brigade, yelling lustily." (Harrington Diary) Perched on his horse and waiting to charge with the 10th Missouri, Captain Hinricks described the scene before him with,

> *Suddenly there was a shot, another, and in a second 10,000 more. The whole country seemed to be alive with demons, such a yell was never heard, but we did not yell long, the next second brought the balls of the enemy by thousands over our head and the shells hurried their way in every direction, having a fiery streak behind them. This was the first time that I ever saw shelling during the nighttime, it is a beautiful but awful spectacle."* (Hinricks, April 16)

According to General Wilson:

> *Just as the troops were ready, the enemy at a short distance opened a heavy fire of musketry, and with a four-gun battery began throwing canister and grape. Generals Upton and Winslow in person directed the movement. The troops dashed forward, opened a*

withering fire from their Spencers, pushed through a slashing and abatis, and pressed the rebel line back to their out-works, supposed at first to be the main line. During all this time, the rebel guns threw out a perfect storm of canister and grape, but without avail. (OR Series 1, Vol. 49, Part 1, p. 363, Wilson's Report)

While the Confederate cannons hurled forth a wall of grape and canister, as General Wilson stated, it was without avail, as the Union soldiers continued to press forward. The Confederate "cannon cockers" were the men of Clanton's Alabama Artillery. Captain Nate Clanton was 30 years old, standing six feet tall, with brown hair, blue eyes, and a commanding presence. While he was an experienced veteran artillery commander, on this night, he made a grave error that would cost the Confederacy greatly.

CSA Captain Nate Clanton

One of Clanton's gunners was Corporal William W. Grant,

who provides a valuable statement as to why the artillery fire was ineffectual. Grant recorded the opening events as witnessed by the Confederate artillerymen on Ingersoll Hill with, "At about 9:30 in the evening of April 16, we heard distinctly the bugle call of the enemy in front of our line to charge. With our six- and twelve- pound brass howitzers, we were ordered to commence firing at an estimated distance of fifteen hundred yards." (CV, April 1915, Vol. 23, p. 164)

It was clearly a tremendous mistake committed by the Confederate artillerymen that allowed the Union attackers to weave their way through the obstacles before them, and assault the Confederate line quickly, and with few casualties. While Captain Clanton made a critical error in ordering his guns set too high, the terrain the Union soldiers traversed provided natural defiles that contributed to their survival as well.

General Wilson described the developing fight with:

The starlight was so faint, however, that nothing could be clearly seen except the flash of firearms. The roar of the artillery and musketry were continuous and appalling, but the enemy fired so high that they did but little harm to our dismounted men. Darkness was their best protection, and, being veterans of four years' experience, they continued their advance unshaken and almost unharmed. (Wilson, p. 261)

For the dismounted Union troops, the quicker they breeched the line, the less fire they would receive as the Confederate artillerymen would not fire into their own ranks.

Sergeant Josiah Conzett of the 5th Iowa one of the first men to breech the Confederate line, and he summed up the initial assault (in his own words) with:

> *Up and away on full run we rushed, our carbines in full play - and they were pumped never before faster... They [Confederates] were protected by strong works - we were in the open, the only thing that saved us I may say was the darkness - our nearness to them and our sudden and quick rush - They overshot us, their guns were not suppressed low enough, their shell and shot as well as their musket or Infantry balls – flew screaming over our heads before they had time to correct their aim. But their rapid fire from fort and Infantry in their entrenchments, lit up the darkness and made the night look bright as day - but it blinded us so that no doubt our aim was bad - but we shot straight ahead and just as fast as we could work our carbines.* (Conzett, p. 77-78)

The Spencer Carbines played a crucial role in the rapid assault of the Union troops. With the ability to fire as they rushed forward, and do so seven times before having to reload their weapon provided the Union troopers with superior lethality and maneuverability.

Union cavalrymen and their accurate, Spencer rifles, in action.

Within fifteen minutes, the exterior line was breeched, and while Union casualties were lighter than expected, nonetheless men were horribly mangled and killed while breeching the line. According to Adjutant Ebenezer Gilpin, as the troops charged the forts, "So near were our men to the batteries that some were made blind by the powder flash." (Gilpin, p. 654)

One of these men that got in that close was Captain Thomas Miller, commander of Company D, Third Iowa, who received a blast of grape shot in his side as he led his men into the Confederate earthworks. Upon the heights overlooking Girard, Alabama, Captain Miller uttered his last words, in which he reflected that it was an honor to die "like a Christian and a soldier." (OR Series 1, Vol. 49, Part 1, p. 493, Noble's Report)

While the 5th Iowa helped drive in the pickets along the exterior line, it was the 3d Iowa that breeched the line and swept across the Confederate earthworks.

On the opposite side of their objective, the cavalrymen reformed their units, as they were ordered not to take any prisoners, but to continue pressing forward. According to Colonel Noble, in this first Union success, his unit "had swept around to the front slightly, and on being halted the new position was a line somewhat oblique to the main line of the enemy." (OR Series 1, Vol. 49, Part 1, p. 493, Noble's Report) This main (interior) line was the long trench running from Ingersoll Hill, now to the Third Iowa's left, and running downhill to the Confederate fort on Red Hill.

While Colonel Noble's unit was halted on the opposite side of the Confederate exterior line, General Upton signaled for Captain McGlasson's detachment to proceed as planned- storm the works, and capture the bridge below. The troopers spurred their mounts, and shot headlong into the breech. In the words of Captain Hinricks:

The order came for us to charge- a sweet thing I can tell you, to charge the enemy's works, when it is dark as pitch and you don't know where and what they are. Particularly sweet when every second your eyes are blinded by the blue light of the shells exploding all around you and passing within four or five feet of the ground across the road you have to travel, not speaking of the grape, canister and small bullets which however likely you cannot see, but only hear whistling by, or striking trees, or something else. (Hinricks, April 16)

The Missouri cavalrymen spurred their mounts and flew through the Confederate line, heading straight down the Summerville Road. Parallel to the road, the Confederates peering out from their trenches undoubtedly saw the approaching cavalrymen, but in the darkness, it was impossible to determine if they were friend or foe. According to one of the Union officers, Captain McGlasson led his men so boldly, "with such coolness and self-possession, indeed, that the enemy supposed his party to belong to their own forces." (Scott, p. 495)

Riding directly across the Confederate line and just above the major fort on Red Hill, McGlasson led his men right on down the road to the 14th Street Bridge. McGlasson coolly halted his column and ordered his men to seize the Confederate guards posted at the head of the bridge.

Lieutenant Frederick Owen and a small detachment of troopers were then dispatched across the bridge to seize the Confederate two-gun battery at the opposite end. While this masterful action was achieved with a bewildering stroke of luck, it did not take long for the Confederates to realize what was happening. For Captain McGlasson and his Missouri troops, their success was bittersweet as they were surrounded by angry Confederates.

Captain McGlasson immediately recalled Lieutenant Owen's

detachment from the opposite side, and under a hail of Confederate gunfire, they raced back from whence they came. Miraculously, McGlasson only lost one man on his incursion behind enemy lines, and while his mission failed, he returned with valuable information.

Realizing that their exterior line had been breached by the Union cavalrymen, all hell broke loss as the Confederates desperately tried to defend their positions. Both Clanton's and Waddell's artillery batteries went into action firing into the area west of Summerville Road. Watching the scene on the opposite side of the river and atop her home on Rose Hill, Ms. Louis Gunby Jones recorded that, "The flashes from the guns looked like streaks of lightning darting from the lower bridge up the river as far as we could see." (Jones, CSU paper)

According to Adjutant William Scott of the 4th Iowa, "From the vast noise they made, it seemed as if the rebels would annihilate the last of their assailants, but all their fire was still too high." (Scott, p.496) Meanwhile, the remainder of the 10th Missouri commanded by Lieutenant Colonel Frederick Benteen had been ordered down the Summerville Road, where they would dismount and try to breech the interior line near the main Confederate fort on Red Hill.

For the men of the 10th Missouri, the attack quickly turned into a hellish nightmare. According to Captain Hinricks:

> *All of a sudden, the column in my front stopped. I heard someone holler, 'there is no road.' Others hollered, 'go to the right, others hollered go to the left.' In short, the company or part of it in my front got tangled up. I turned to where the most went, to the right, the company in my rear charged in on me. All got tangled up, to add to the confusion it happened that a shell fell among us killing 2 horses and one man. At the same time, a tremendous fire was opened in our*

right with small arms... the bullets were coming thick and fast, the enemy now commenced from two forts to treat us with shell and canister. 3 horses of my company were killed and 2 men struck with canister. They were coming by the basket full.

Several officers asked me to take the men away from there and reform. I did so and sent several orderlies to find the Colonel and inform him of our whereabouts. We tried to retrace our steps, but we were so mixed up that we took a road leading direct to the fort of the enemy. (Hinricks, April 16)

This confusing situation is especially ironic in that the 10th Missouri Cavalry, United States Army was probably receiving fire from the 10th Missouri Artillery, of the Confederate States Army. Yet, it did not take long for Captain Hinricks to collect his wits and try to react wisely to his predicament.

Hinricks continues his experience stating that the Confederate guns on Red Hill:

Sent a ½ dozen shells whizzing over our heads. We were within 60 yards of the fort, and had I but known the situation at that time, I should not have hesitated a moment to attack even without orders. We reformed again and were halted within less than 125 steps of the muzzles of the rebel guns. We were hardly reformed when a solid shot struck right between us.

The men stood. A second shell struck and exploded within 4 feet of my horse, covering myself & my company with mud. Most of the men broke & as I could not talk to them for fear that the enemy would send us more such customers. I found myself with ½ dozen men.

> *The Colonel passed and when I reported the fact that several men were wounded and horses killed, he made use of the expression, "I'll give a god damn if the last god damned one got killed. I think the gentleman don't know it all yet himself, or he would not form men 125 yards from a battery what had the range as it had it particularly when there is no need for it. My men all wished him a speedy journey to hell.* (Hinricks, April 16)

Simultaneously, uphill (north) of the 10th Missouri, the Third Iowa and the two companies of the Fifth Iowa, which had rallied on the far slope of the Confederate exterior line, were ordered to make a left wheel and swing around to attack the interior line along Summerville Road.

As the Third and Fifth Iowa regiments moved out, they descended into a wooded ravine and marshy area, crossing a brook and "passing in the profound darkness over fences, ditches, and sloughs, with no other guide than the light and roar of the rebels' fire." (OR Series 1, Vol. 49, Part 1, p. 494, Noble's Report)

Scrambling as fast as they could back up the opposite side of the ravine, the main trench line slowly came into view. All the while, the Confederate batteries on Ingersoll Hill and Red Hill fired continuously into the trees above them. With crashing tree limbs and shell fragments raining down on them, the Union troops then had to weave their way through the 100-yard web of abatis and slashings just to get in to position to attack.

The movement was extremely difficult, but three companies of the Third Iowa made their way to the trench line, and "with great bravery seized the nearest angle of rifle-pits and held it obstinately, driving off or capturing the defenders." (Scott, p. 496) But above just above these companies, their comrades attempted to charge the two-gun lunette on the western slope of

Ingersoll Hill, and were forced to retreat.

Unable to maintain their forward momentum of the assault, the Iowans that could not take the lunette, straggled back to the Summerville Road. The portion of the Third that had gained a toehold held out as best as they could, but they could not press their breech without reinforcements.

From where the Union generals observed the action, it looked as if the Third Iowa was being cut to pieces, so Colonel Winslow ordered seven companies of the Fourth Iowa to dismount and assist in the attack towards against the Confederate interior line along Summerville Road. For the assault, the dismounted troops of the Fourth Iowa were organized by battalions.

Captain Lot Abraham commanded the 1st Battalion of the 4th Iowa (companies A, D, and K), while Captain Newel B. Dana commanded the 2nd Battalion (companies C, F, I, and L). (Scott, p. 496)

The Third battalion of the Fourth Iowa remained mounted, and was prepared for a cavalry charge. The two dismounted columns rushed forward and formed to the south, below the positions being attacked by the Third Iowa.

Sergeant Harrington of the 4th Iowa noted that as they moved into position, "We were in short range of the forts, which threw shot and shell, grape and canister, thick and fast. Fortunately for us, it was so dark, the Rebs could not see us and shot too high." (Harrington Diary)

Thus, looking eastward towards the Summerville Road and the Confederate trench line, from left to right (north to south) the Third Battalion (four companies) of the Fourth Iowa, in reserve, was mounted and standing by in the Summerville Road, at the top of the heights and adjacent to the Confederate earthworks on the exterior line. Several hundred yards below them were the Third Iowa (six companies) and the two companies of the 5th Iowa.

Beside them and on their right, were the two dismounted battalions of the Fourth Iowa, and on downhill in the extreme right was the 10th Missouri.

At this point, the Third Battalion of the Fourth Iowa being held in reserve for a cavalry charge into wherever the line could be breeched was the only unit of the 1st brigade that was not engaged. The remainder of Winslow's 1st Brigade was on line and attempting to breech the main Confederate trench line along Summerville Road.

According to Adjutant Gilpin of the 4th Iowa, "A wild exultation seized the soldiers, and I believe our division could have whipped anything in the Confederacy. It was grand to see and hear the battle at night- all dark except when the scene was illuminated by flashes of the guns and glaring brilliancy of volleys from forts and rifle pits." (Gilpin, p. 654)

The Confederates in the trench line beyond Summerville Road deployed skirmishers forward to meet the Union troops before them. According to Private Charles K. Henderson:

> *As we leaped over our breastworks or trench, I said, 'They will get me now.' The Federals raised their battle cry. After a moment, we were ordered: 'Fall to the trenches.' We were soon in. Henry Greene of Chattahoochee County stood at my left, a brave boy. He said to me: 'If I get killed, I want you to take care of me.' I said the same to him.'* (Barfield, p. 752)

Opposing these young inexperienced Confederates were the best men the Union could muster, and their battle cry, was "Selma!" as they charged the line before them. Again,

Private Henderson describes the scene in his own words with:

> *The battle raged. I had a Mississippi Yager to shoot with. I heard the minie balls Sh-Sh, hiss near my ears. And the clanking of the swords of the Federal cavalry in front of me. The whole battle line of the Confederates popped like firecrackers. The battery cut off the tops of many pines in front of me. The federal infantry crawled to our trenches. I heard one federal call to a Confederate: 'Throw down your gun and surrender.'* (Barfield, p. 752)

According to Adjutant Scott of the Fourth Iowa, "The obstinate possession of the Third and this new assault were too much for the enemy. They broke from all the works in the vicinity of the lunette." (Scott, p. 497)

Privates Henderson and Greene saw an officer in front of them yelling to "fire oblique to the right." Greene excitedly pointed and said to Henderson, "That's a Yankee!" Then, "A company broke on our right. All was confusion now." (Barfield, p. 752) Private Henry Nuckolls of the 2nd Georgia State Line had rushed all the way from his family farm in Forsyth County, Georgia to fight the Yankees. This battle was his first, and although he had just turned nineteen, it was also his last.

The Confederates took numerous casualties as the Iowans pushed forward firing their rapid-fire Spencers as fast as they could. The Confederate firepower was rapidly overmatched as they struggled with their awkward and antiquated muzzle-loaders. The area just below the 2-gun lunette was repeatedly hammered and Major Nisbet's 26th Georgia Battalion with reserves on either side of his position worked hard to bolster the line. According to Major Nisbet, "We repulsed the charge on our front, but the 'citizens' gave way." (Nisbet, p. 255)

As the 4th Iowa breeched the line, they rushed amongst the Confederates firing wildly into the ranks of the dismayed and inexperienced Confederate reserves. Adjutant Gilpin noted that, "The Confederates held stubbornly to their guns until our boys were in among them and forced them to surrender." (Gilpin, p. 654)

According to Private Alphonza J. Jackson of Company G, 2nd Georgia State Line, "We then realized our situation. We, of course, leaped out of our ditches and made for the bridges." (Alphonza Jackson, Diary and Letters. Box 76-7, Microfilm Library, Georgia Department of Archives and History, quoted in Bragg, p. 107) This was the greatest fear of the Confederate officers, and as men saw their comrades peeling out of the line and running for the bridge, they too, fled the field in a panic.

The Union cavalrymen yelled after them, "Surrender! Throw down your arms!" But young men who honestly believed they were running for their lives could not be urged into submission. Private Jackson continues his description of flight with:

While running down a steep slant, I remember running into Lieutenant Colonel Evans and knocking off his hat. I never stopped running through a field in the darkness, I ran into a gully over my head and had to go down it until I found a gully leading out from it before I could get out. We finally came to a fence. I remember I climbed to the top of it and being almost out of breath, just fell off. Then up the railroad embankment and crossed the river on the railroad bridge. Once across, I felt safe. Some soldiers had piled up bales of cotton there for breastworks and begged us to rally, but we were so demoralized and scattered that we didn't. (Alphonza Jackson, Diary and Letters. Box 76-7, Microfilm Library, Georgia Department of Archives and History, quoted in Bragg,

p. 107)

Try as he may, Lieutenant Colonel Beverly Owens, deputy commander of the 2nd Georgia State Line could not restrain his men from flight. The Confederate line had broken; however, the fight was not over. According to Colonel Noble, the Third Iowa maintained their position while the Fourth Iowa turned to the right and formed to charge the fort on Red Hill.

Colonel Noble reported that, "Considerable portions of Companies A, B, and I, under Captains Wilson, McKee, and Arnim, took captive the rebels at an intermediate point of the entrenchments, seizing the garrison flag of the post, Sergeant Birdsall, Company B, gaining this trophy… Private Tibbets, Company I, captured the battle-flag of Austin's battery [battalion] in this assault." (OR Series 1, Vol. 49, Part 1, p. 495, Noble's Report)

Mistakenly referred to as a battery, Austin's 14th Louisiana Battalion was a small group of sharpshooters, employed as cannon guards.

Captain Lot Abraham

Private Andrew W. Tibbets

They had followed their commander, Major John E. Austin throughout the war, and although their ranks had dwindled to less than thirty men, they were experienced veterans and accurate riflemen. Colonel Von Zinken had known these men since their inception, and if he had more of them, the battle may have ended differently.

Austin and Von Zinken had served together in the Army of Tennessee, they were both from Louisiana, and consequently, they shared an honorable and blood stained heritage. (Sifakis, Bergeron, CS records)

During this part of the battle, the Union suffered its greatest casualties, and the Third Iowa earned the unenviable position of having the most. The two Confederate guns in the lunette were a section belonging to Clanton's battery and after they were taken, Lieutenant Forker led Company B up the crest of Ingersoll Hill to charge the main position from behind. From his fort on

Ingersoll Hill, Nate Clanton had watched in disbelief as the panic-stricken Confederate main line below him retreated from the field.

By the time, Lieutenant William Forker reached the fort, according to Corporal William Grant, Clanton had already ordered "the sad and what turned out to be the last order to take care of ourselves," and they too fled the field. Very few of Clanton's men were captured that night, as Corporal Grant recorded, after spiking the guns, "We left the Girard hills in squads of six to twelve." (CV, April 1915, Vol. 23, p. 164)

Meanwhile, the southern end of the battlefield turned into a giant brawl and a mass of confusing violence. After the surprising charge of the 10th Missouri down to the bridge, the Confederates lit several houses on fire near the bridge to illuminate it. While the men could see one another, in the eerie glow of the fire, and the contrasting darkness, it was still difficult to distinguish friend from foe.

General Upton joined his men as the pressed the fight, and yelling above the din, he urged his men forward, yelling, "Charge 'em! Charge em!" According to General Wilson, "It was in this movement that Upton displayed the extraordinary insensibility to danger which always characterized him." (Wilson, p. 263)

Adjutant Scott of the Fourth Iowa relates that, Both General Upton and Colonel Winslow were urging their men forward and yelling, Selma! Selma! Go for the bridge!" Take no prisoners!" Go for the bridge!" (Scott, p. 497) While many men acted heroically that night, Captain Lot Abraham after leading his battalion across the main trench line, wheeled it to the right, and faced off against the strongest position on the line, the fort on Red Hill.

Again, Captain Abraham had Captain Dana's battalion on his flank, and together, they focused their efforts in another major thrust into the Confederate line. This time, they rushed the battery that lined the street in front of the fort's north wall.

Captain Abraham's battalion bore the brunt of the fight, and he led companies A, D, and K in a hand-to-hand battle with Waddell's artillerymen. Private John Kinney of Company L, 4th Iowa captured the standard and bearer of the 10th Missouri Battery, and according to Kinney, "I had a tussle with the fellow to get the flag." (OR Series 1, Vol. 49, p. 399)

While some of the cavalrymen wrestled with the gunners, others scaled the ramparts of the fort. After two attempts to breech the fort's north wall, the Union troopers swept through the Confederate bastion, and silenced the Confederate guns forever.

Thus, Waddell's Artillery was the Confederacy's last battery to silence their guns. Searing from the heat of a long day's battle, the gun barrels would cool long before the men that fired them. Forced into submission by an overwhelming force, men like Captain Richard Bellamy and his younger brother, Sergeant William Bellamy had both fought a long and bitter fight. Except for the elder's long flowing beard, William was the mirror image of his elder brother, as both were tall, with dark hair and ice blue eyes.

For the last two years, William had watched his brother intently and like Richard, he too had become a skilled artillerist; however, these artillerymen stood no chance against the brutal close-fight delivered by Captain Abraham and his Iowa warriors.

For the Bellamy crew, the night marked the end of their military careers, as the brothers were corralled amongst the many Confederate prisoners of war. Amidst this chaotic scene, some men did avoid capture and they slipped away.

Private James B. Grant of Bellamy's Battery did like his older brother Corporal William Grant of Clanton's Battery, and he too escaped into the darkness. (CV, April 1915, Vol. 23, p. 164)

While some Confederates disappeared into the night, many of them would never move again. One of these men was Captain Isidore Guillet, Colonel Von Zinken's Adjutant. Killed upon his horse, the young man from Louisiana died attempting to rally his men on Red Hill. Like his two brothers before him, he too gave his life for the Confederacy.

In a final attempt to quell the Union assault, Buford's small detachment of cavalry plunged headlong into the fray with a desperate cavalry charge. In this fight, another Confederate casualty in the struggle on Red Hill was Captain John S. Pemberton.

According to his friend and fellow cavalryman James D. Carter, as they charged the mass of Union troopers pouring onto Red Hill, Pemberton suddenly reeled in his saddle.

A Union sharpshooter found his mark and Pemberton was plugged directly in his chest. As he reeled in the saddle, and struggled to stay on his mount, another Union soldier slashed Pemberton with his sword. Carter rushed to his aid, grabbed the horse's bit, and hauled his friend to safety. Another Confederate cavalryman, Private George Turner was shot in the leg and his horse was killed beneath him.

Another Confederate cavalryman was killed and his face was so horribly mangled his name was lost to the ages. This trooper is buried in Phenix City beneath a CSA marker that states "Unknown Confederate Cavalryman." The ferocity and momentum of the Union thrust combined with the accurate fire of their carbines was simply too much for the Confederate cavalry who suffered fifteen casualties.

 Realizing that the charge was too little, too late, and unable to stop the Union onslaught, General Buford chose to retreat. Thus, Generals Cobb, Buford, and Adams, along with Colonel Von Zinken, and what remained of their staff, wheeled about and retreated from the battlefield. Joining their fellow Confederates in the mad dash to cross the 14th Street Bridge, their enemy also followed in hot pursuit.

 After subduing the fort on Red Hill, Abraham left a guard on the captured fort, and like the retreating Confederates, he and Captain Dana led their units into the mob heading into the 14th Street Bridge. Racing down the hill towards the bridge, the troops of both sides were intermingled, thus the Confederate battery posted on the opposite bank dare not fire into the mass of men rushing over the bridge.

 Following them were the mounted columns of the 10th Missouri and the Third Battalion of the 4th Iowa, with Colonel Winslow at the head of the column.

 Adding to the confusion, the bridge was several hundred yards long, and a covered two-lane wagon bridge, thus the numerous men that jockeyed for position within its dark corridor could not possibly tell who was who. Inside the bridge, the pungent odor of turpentine filled the air. In fact, like the other bridges, the 14th Street Bridge was saturated with flammables, and its joints were filled with cotton, which had been soaked with turpentine as well.

According to Private Washington Crumpton, a fleeing member of the Georgia militia, the bridge was more like a tunnel, which he described with:

> *In the dark covered bridge, crowded with men and horses, there was danger of a footman being knocked down and tramped to death, so instinctively, every fellow reached out his hand and steadied himself on a comrade. When the opening of the bridge, on the Columbus side was reached, it was light enough for me to be impressed at sight of the dark clothes my companion wore. I thought I observed others with dark clothes- The truth was, some of the Yanks had come over the bridge with us.* (Crumpton, p. 100)

As the intermingled forces burst out of the bridge and poured into Columbus, they were immediately confronted with the Confederate two-gun battery.

At the front of the mob, fleeing for his life was Private Crumpton, and according to him, "At the mouth of the bridge, a drunken Confederate Cavalry General sat on a horse, pistol in hand, swearing he'd shoot the first man who passed. They were passing in a stream, but he was too drunk to know what was going on. On either side of him were a cannon and an officer threatening to fire into the bridge, jammed full of men." (Crumpton, p. 100-101)

While the gunners hesitated to unleash their grape and canister into the mixed mob of friend and foe, one Confederate soldier attempted to set the bridge on fire, but he was quickly "crushed in the act by a clubbed carbine in the hands of a man of Company K." (Scott, p. 499)

Whether the Confederate Cavalry General mentioned by Private Crumpton was General Buford is unknown, but Colonel Charles A.L. Lamar of General Cobb's staff was there as he quickly engaged the Union troops. Lieutenant William N. Martin, the elder Confederate of the Alabama Martin clan recorded that Colonel Lamar "came galloping down the street, sword in hand, calling on his troops to rally and hold the bridge." (Russell County History, p. C-46)

A firefight ensued and the Confederate artillerymen were shot where they stood behind their cannons. Several Confederate soldiers fired small arms into the Union cavalrymen wounding two, and killing Sergeant Joseph H. Jones. In this firefight, several other Confederates were also killed including Private Alexander Robinson, a local factory worker turned soldier.

At the same time, Adjutant William Scott of the Fourth Iowa on horseback was attempting to subdue Lieutenant William N. Martin who was on foot. As Colonel Lamar rode by the struggling men, Adjutant Scott spurred his horse, knocking down Martin, while simultaneously drawing his pistol and killing Colonel Lamar.

While Adjutant Scott checked the Colonel Lamar to make certain he was dead, William Martin made his escape, and slipped into the countryside. (Russell County History, p. C-46; CE June 27, 1865)

While the fight at the bridge was being waged, Private Robert C. Wood of Company A, 4th Iowa Cavalry being held captive right next to the bridge in the Mott house, called for his fellow troopers to help him. A few men from companies A and I helped Private Wood capture his captor, Colonel James C. Cole. (OR Series 1, Vol. 49, Part 1, p. 499)

Unbeknownst to these men, Private Crumpton was trying desperately to free himself from the iron picket fence, which surrounded the Mott House. After pushing his way through the firefight at the mouth of the bridge, he ran down to the left of 14th Street and tried to enter the grounds of the Mott House, but the gate at that point was locked.

With the Yankees helping Private Wood escape from inside the Mott House, Crumpton panicked and in his own words he explains that:

> *I shoved my gun through, and mounted the fence. In attempting to let myself down from the top, a sharp iron picket passed through the seat of my new jeans pants and there I was, impaled between the heavens and the earth.*
>
> *The Yanks had set fire to a house on the other side, making it as light as day, and were firing across the river at the fellows who were fleeing through the spacious grounds.*
>
> *To be killed on top of that fence was too horrible to contemplate, so I put my heels against the railing and grasped the top of the rods with both my hands, pushed backwards and with a crash the seat of my new breeches gave way and I came down face foremost to the ground.*
>
> *As I came out into the street on the other side, I found the Yankee Cavalry had crossed the river above and were passing into the city, so I took an outing in the woods for two weeks with a squad of escapees.* (Crumpton, p. 101)

While Private Crumpton succeeded in making his escape, other Confederates continued a bitter resistance in the streets of Columbus.

The mounted columns of the 10th Missouri and the Third Battalion of the Fourth Iowa spread into the streets where they ran down their enemy. According to Sergeant Harrington of the 4th Iowa, "We found the Rebs thick, and greatly surprised at our order for them to surrender. Some of them had to be shot first." (Harrington Diary)

Sergeant Norman Bates won the Medal of Honor for capturing the flag of the CSA 7th Alabama Cavalry

Several skirmishes occurred east of the 14th Street Bridge, but the Confederates were quickly routed. In these final desperate exchanges, several more citizen/ soldiers were killed in the streets, including J.J. Jones, Evan Jones, William Smith, and at least two other men from Columbus.

In short order, pockets of Confederate resistance were quelled, and a detachment of Union cavalry rushed towards the train depot. Yet, "not knowing the way… a considerable train had got away toward Macon, filled with officers, soldiers, and citizens of position." (Scott, p. 500)

The escaping Confederates included Major General

Howell Cobb, Georgia Militia Chief of Staff Robert Toombs, and 600 men of the Georgia State Line. Both General Adams and General Buford raced around the city on horseback and headed southward to avoid the Union patrols, which rapidly spread throughout the streets of Columbus. (Scott, p. 500; OR Series 1, Vol. 49, Part II, p.383)

Except for the few "odd angry shots," inherent in mop-up operations, the battle was over by 10:00 pm. Sergeant Harrington patrolled the streets of Columbus searching for "Rebs," and in his diary, he wrote:

We rode through the town and captured more than 5 times our number. It is now 1 AM, and we have just stopped at the RR Depot. I am tired but not hungry, for I have had plenty to eat tonight. I praise God for his care, and that he has given us the victory so easily. (Harrington Diary)

According to Captain Hinricks of the 10th Missouri, "We were master of the city. The rebels surrendered by the hundreds as our dismounted men had done considerable exertion and many a dead rebel covered the ground." (Hinricks, April 10) Adjutant Scott of the 4th Iowa recorded his observations that night with:

There was no more noise, except the occasional cheers of the victors. The burning buildings in Girard, which the rebels had fired to light their operations, continued to cast a lurid glow upon the scene of the conflict. The victory was perfect. (Scott, p. 500)

From the Confederate perspective, a summation of the battle on Easter Sunday, April 16, 1865, was recorded in the final entry of the logbook at the Columbus Naval Iron Works. It states,

Heavy firing at upper bridge until between 9 & 10. When the enemy broke thru our lines & obtained possession of the bridge & the City. Our forces leaving as rapidly as possible... FINIS. (Turner, p. 233)

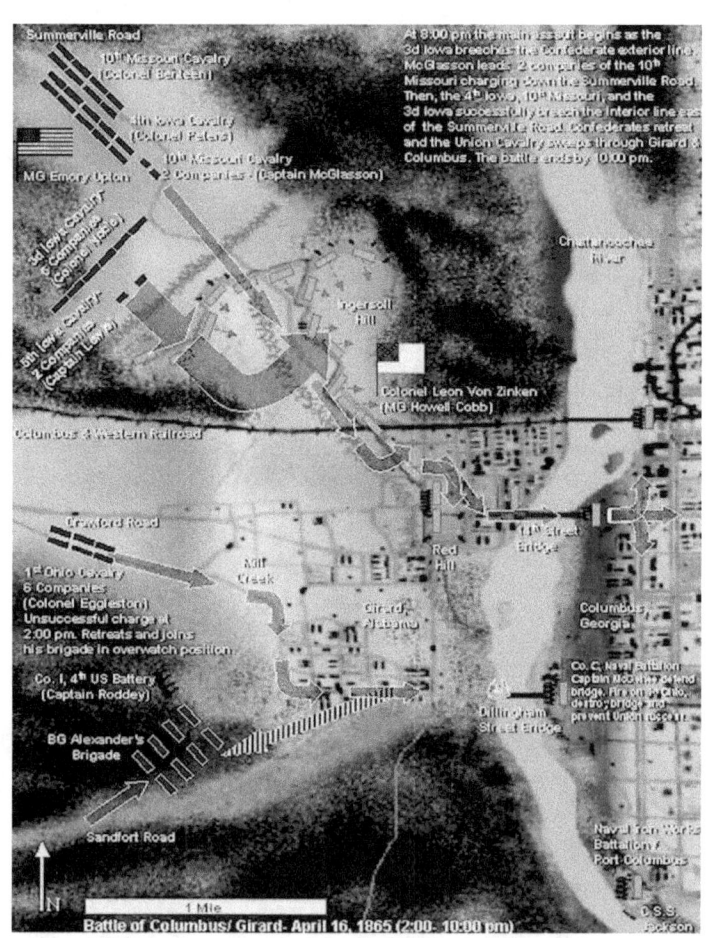

Wilson's assault on Columbus/ Girard

Order of Battle

Forces engaged in the Battle of Columbus/ Girard:

Union:
Commander- Brevet Major General James H. Wilson, Cavalry Corps

4th Division- Brevet Major General Emory Upton.

1st. Brigade. Brevet Brigadier General Edward F. Winslow.

    3rd Iowa, Colonel John W. Noble.
    4th Iowa, Lieutenant Colonel John H. Peters.
    10th Missouri, Lieutenant Colonel Frederick W. Benteen.

2nd Brigade. Brevet Brigadier General Andrew J. Alexander.

    5th Iowa, Colonel J. Morris Young.
    1st Ohio, Colonel Beroth B. Eggleston.
    7th Ohio, Colonel Israel Garrard.
    4th United States Artillery, Battery I, Lieutenant George B. Rodney.

Confederate: (estimated 3,250 men and 24 cannons)
Commander- Major General Howell Cobb, Georgia State Reserves

1st & 2nd Georgia Infantry and miscellaneous troops- (1,500 men)

General Abraham Buford's 7th Alabama Cavalry (remnants-estimated 350 men)
Colonel Leon Von Zinken (overall commander)

Local Defense Force, Columbus, Georgia (estimated 1,100 men)

Local Defense Force, Girard, Alabama (estimated 100 men)

Clanton's Alabama Artillery Battery (estimated 100 men and 10 guns)

Waddell's Alabama Artillery Battery (estimated 100 men and 10 guns)

CHAPTER 6
THE AFTERMATH

By 11:00 pm on the evening of April 16, 1865, the fighting in Columbus and Girard had ended. Adjutant Gilpin of the 4th Iowa wrote, "Now that the battle is over, and we have possession of the city, strict discipline is enforced. Contrasted with the night we took Selma, it seems very quiet." (Gilpin, p. 654)

If not for the glow of burning structures and the moans of the wounded, a certain sense of peace had returned to the Chattahoochee Valley. Yet, the night air was cold, and in the blood-soaked hills of Girard, shattered men, both Union and Confederate struggled to survive.

Across the river in Columbus, the men and women working at Walker CSA Hospital did what they could to save them. One of the Union officers recorded that his unit entered Columbus around midnight, and "until morning we could hear the slow rumbling of ambulances to the hospital, where the surgeons were busy." (Gilpin, p. 654)

Columbus had become a major CSA treatment center after the fall of Atlanta, and at its peak, Columbus housed 10 CSA hospitals and convalescence centers. Walker CSA Hospital was formerly a large hotel on the corner of 10th and Broad Streets, and it was the largest and best facility in Columbus.

Dr. Carlisle Terry was the Chief Surgeon in Columbus, and throughout his tenure, he fought with Confederate administrators to better fund the hospitals. He even bought food for the soldiers with his own money and Dr. Terry actively petitioned the local government for additional funding. Working primarily with the ladies of Columbus through their *Soldiers' Aid Society*, Dr. Terry participated in fund-raiser events, gathered blankets, and obtained badly needed supplies for the soldiers.

Dr. Terry was an advocate of the importance of diet, and coupled with his skills as a surgeon, the soldiers were in highly capable hands. Both Union and Confederate surgeons worked diligently through the night trying to save lives, and a total of 178 men were treated that night. (OR Series 4, Vol. 3, pp. 718-20; Columbus Hospital Records)

The Walker CSA hospital log records that on the night of April 16th, 144 Confederate soldiers, and 35 Union soldiers were treated for their wounds. These records also indicate that at least 43 Confederate casualties experienced the lethality of the rapid-fire Spencer carbine that night. In contrast, the Confederate minie balls contributed to only 17 Union casualties.

Additionally, several soldiers that were treated that night, died later because of their wounds, and in this manner, at least 17 Union and 4 Confederate soldiers died between April 16th and July 14, 1865. (Muscogee County Hospital list)

No Confederate reports of the battle were ever submitted as for them, the war was over. The Union reports indicate casualty figures for their own men, and General Upton listed these as 6 killed and 33 wounded, while the Chief Surgeon Francis Salter, stated in his report that, "We lost in killed and wounded 39 men, of whom 7 were killed."

When Wilson's Cavalry Corps departed Columbus, Surgeon Salter detailed his assistant, Surgeon Samuel Whitten of the Third Iowa to remain with 35 Union patients at Columbus. (OR Series 1, Vol. 49, Part 1, p 407-409, Salter's Report)

The Muscogee County Sexton (Coroner) reports indicate that by the end of April, seventeen of the thirty-five Union patients died. Sexton R. T. Simons recorded, "I buried 17 federals who were killed and died of wounds rec'd in the affair of 16 April." (Columbus Sexton's Reports, Book E, 1853-1866, City Council records, City Clerk's Office)

Through the years, the remains of these men were re-interred, with some forwarded to Iowa cemeteries, and at least 7 of them are now buried in Andersonville National Cemetery.

Throughout the night and into the early morning hours of April 17, Union troops patrolled the streets and byways of Columbus searching for Confederate troops. One of these Union patrols observed a man departing the downtown area in a carriage and traveling at a rapid pace.

Pursuing the shadowy figure along Hamilton Avenue, the Union patrol warned him to halt, but the carriage continued moving rapidly down the street so they opened fire. Despite repeated warnings to halt and a hail of gunfire, the determined man escaped them and the hapless Union patrol moved on.

Unbeknownst to the soldiers, they had nearly killed Mr. William Young, the prominent president of the Columbus Bank and the owner of the Eagle Mill. Mr. Young had remained at his office in the mill until he felt it was safe to make a run for his home.

Mr. Young would probably have stopped for the patrol, but the elderly gentleman was nearly deaf and he simply could not hear their repeated warnings to halt. While Mr. Young was shot several times in his encounter with the Union patrol, he reached his home and ultimately survived the incident.

Unlike Mr. Young, several civilians were tragically killed that night in downtown Columbus, but in the darkness and bedlam of battle, the Union soldiers had a very difficult task of determining when to withhold their fire. While some Confederate soldiers surrendered, many of them turned fugitive and made harrowing escapes as they fled into the surrounding countryside.

Like many of his fellow comrades, Private William Grant of Clanton's battery lived within a day's journey of Columbus, so he simply fled for home at Uchee, near Crawford, Alabama. Traveling westward with twelve of his comrades, Grant recalled that:

We tramped over hills and through the woods, giving the public highways a wide berth to avoid capture… Knowing that clothing was scarce at home, I had put on an extra jacket and trousers, which had recently been issued, and we stopped to rest I was perspiring freely. Soon I was so cold that I could not sleep. At sunrise, we resumed our weary march, stopping now and then for a bite of cornbread and some buttermilk. (CV, April 1915, Vol. 23, p. 164)

Privates Charles Henderson and Henry Greene, who had made a pact while fighting in the trenches above Girard, also fled the battlefield together. Upon reaching Linwood Cemetery in Columbus, about one mile northeast of the 14th Street Bridge, the young men split up as they continued onto their homes in neighboring counties.

Henderson later recalled, "I have not seen Henry since that night. I got home to breakfast and surprised Pa and Ma. I said: I have just come from a battle in Columbus." (Barfield, p. 752)

According to the records of the Confederate Naval Works Battalion, three of their men fled the city via a land route and were promptly captured, but those who escaped down river fared quite well. While the gunboat, *CSS Jackson* was left in port, the other vessels made it to a point near Race Pass on the Chattahoochee River, twelve miles below Columbus.

At Race Pass, the Confederate sailors doused the *CSS*

Chattahoochee with kerosene, and destroyed it by fire. According to Lieutenant William Carnes, "there seemed little use for Navy officers" and after the *Chattahoochee* was destroyed, he traveled west in search of Forrest's cavalry. (Navy Grey, p. 234)

Buford's Cavalry was also in search of Forrest's Cavalry, as they had escaped during the night and as planned they were to meet at a predetermined rally point. In addition to Buford's detachments, General Adams and Colonel Von Zinken were also attempting to re-cross the Chattahoochee in hopes of regrouping with General Richard Taylor's forces. Yet, cut off from communications, these men had no idea how thoroughly the Union had shattered Alabama.

Moreover, the Confederacy was nearly ended, but again, they were totally ignorant of the situation beyond the Chattahoochee Valley. Unfortunately for Buford's cavalry detachments, the selected rally point was West Point, Georgia, which unbeknownst to them, was also occupied by Union forces.

Meanwhile, Buford had escaped to the south, thus to reach the rally point, his detachments had to ride all the way around to the east and north in a giant arc to reach West Point, Georgia. As Buford's men crossed the railroad between Columbus and Macon, Union cavalry patrols fired upon them, and they narrowly escaped.

Again, as Buford's detachment crossed Macon Road, Union patrols engaged them and gave pursuit. As they approached West Point, Wilson's men were descending upon them by that route, thus the cavalrymen split up into smaller parties to prevent capture. During the next two days, Buford's men hid in the woods between Columbus and West Point.

While about one-half of the roughly 3,250 Confederate

combatants escaped the battlefield, according to General Wilson, and other Union officers, there were at least 1,500 prisoners that either laid down their arms, or were forcibly captured. (Wilson, p. 265)

On the night of April 16th, Captain Hinricks of the 10th Missouri was assigned to patrol the main route used by the retreating Confederates, Macon Road. After posting his guards for the night Hinricks entered a local home where he found a wounded Confederate. While the soldier explained that he had been shot by his own officers for refusing to fight, more likely, the soldier was looking for sympathy lest he be taken captive.

Hinricks got little rest that night as "several rebs came in and gave up," and early the next morning he awoke to the sound of his pickets firing. Hurrying out of bed, Hinricks learned that about twelve Confederates had been spotted down the road and they had fired on them, so the pickets returned the fire.

Hinricks and three of his men gave chase and captured three of them, and noted that the road was filled with enemy caissons, artillery and other valuables. In the light of day, Hinricks began to survey the destruction that had been wrought during the night. (Hinricks Diary)

Hinricks recorded his observations with:

I passed through the streets and noticed our army of Negroes which followed us and which numbered by the thousands helping themselves from piles of confederate clothing to full suits of gray. They were marched along by 4s and supplied as they passed by. At another they were helped to a gun and a filled cartridge box, shoes, caps, and underclothes. The poor citizens also were helping themselves freely to everything. (Hinricks Diary, April 17)

While the scene in Columbus seemed bizarre to Captain

Hinricks, a virtual army of freed blacks had followed Wilson's corps through Alabama and on into Georgia. Unable to turn them away or outpace them, General Wilson accepted them with pity, and provided them with what food and clothing his corps captured along the way. Captain Hinricks continues his observations with:

I re-crossed the bridge to visit the battleground and examine the enemy's defenses. I must say that when I arrived at the place where we had been for ½ an hour under the crossfire of 2 batteries I was astonished that we had been successful at all, for it was not over 125 steps to the nearest battery and we had to run the gauntlet for ½ mile within 75 yards of the enemy's rifle pits, which ran parallel with the road to the bridge.

The trees bore witness of the terrific storm of shot and shell, which had been hurled up against us as almost every tree was spotted. Our dead had been already gathered, but the rebs were yet laying about in the works and they were at work gathering the wounded.

Thousands of guns layed about and testified that the struggle had been a hard one. I met a lot of women and children who had been caught in the storm last night. Several rebs had been killed in their yard, they layed there yet, and their poor folks had heard that Forrest would attack us tonight, so they had decided to leave their home and seek refuge in the country- they did look pitiful indeed. I convinced them to return to their homes, as there was no danger of Forrest attacking the place. They listened and returned with me.

They told me that the rebs had taken away their last crumbs of bread and now that we had burned all the factories, they did not know how to procure food for

> *their starving children; just then I noticed 10 or 12 Negroes running from town, each with a bag of confederate meal on their shoulder. I made every one of them turn back and carry their loads to the homes of those poor women. The city was now burning on every corner and the explosions of powder and shells could be heard every second.* (Hinricks Diary, April 17)

Adjutant Gilpin of the 4th Iowa also recorded his observations of the "morning after" with, "Our headquarters are at the 'Battle House.' Up early and out in the city. The forts are full of prisoners. Prisoners and artillery everywhere… We are Masters of the situation. (Gilpin, p. 654) General Wilson had the prisoners collected and placed in a makeshift holding pen "in a factory yard near the river." (CE June 1, 1865; CE November 30, 1938; OR Series 1, Vol. 49, Part I, p. 478)

According to numerous reports, the prisoners consisted of "boys from eleven to fourteen years old, and men so old that they could scarcely hold guns." (Telfair, p. 137) There were however, many seasoned veterans amongst the old men and boys held as prisoners of war.

One of these men was Major John Nisbet who provides an interesting account of his captivity after fighting the "last battle" the night before with:

> *The next morning, I awoke in the camp of the Yankee provost guard. A tall cavalryman called out to me: 'Hello, Major Nisbet!' I went over to the guard line and asked: 'Do you know me?' He answered: 'Don't you remember taking a batch of prisoners in 1864, from Sherman's army to Andersonville? I was one of those fellers.*
>
> *We all agreed that you treated us well for prisoners, and now I am going to do what I can for you.' He went*

away, soon returned with a good breakfast. (Nisbet, p 255-56)

Several unusual incidents occurred on the "morning after," and one concerns the "Battle House" that Adjutant Gilpin mentioned as being the Union headquarters. This was the same house that played a prominent role in the battle, the night before, and it is known as the "Mott House."

Owned by the ardent Unionist, Randolph Mott, it is an elegant home and it still stands to this day. Its elegant staircase was modeled from a palace in Milan, and in 1851, its first owner Daniel Griffin had the towers and brick wall that surrounds the property constructed. He sold the house in 1857 to Mr. Mott, because his wife was bothered by the constant dampness that prevailed so close to the river.

Mr. Charles Swift wrote an interesting article concerning General Wilson and Mr. Mott that states in part:

With an originality and independence that was characteristic of him, Col. Mott was an ardent and outspoken Union man all during the war, but his helpfulness to those in need and whose main supports were the Confederate soldiers, fighting and falling in the first ranks of the Southern cause, was an immunity to his unrepressed "freedom of speech," which he spoke in a spirit of triumph when he invited General Wilson to be his guest.

He and Andrew Johnson were close friends as young men in North Carolina before the latter came to Georgia and the former moved to Tennessee. The southwest corner of the Mott lot abutted about opposite to where Col. Lamar was killed by the last shot of the "closing scene," except perhaps, the one that killed the noble and much regretted young Alexander W. Robinson.

After the last stand at the Columbus end of the 14th Street bridge, and the enemy, probably to show inhabitants that the city had been captured, did firing on Broad Street. Col. Mott's body servant at the time was a young, active, good looking and proud-stepping African Negro, named Frank Barnbush. He had been one of the cargo of the slaver "Wanderer," and it is very likely that Col. Mott bought him from Col. Lamar.

Frank probably little understood the meaning of the noise and firing in this last excitement, which he might have conceived as something like the ambuscade in the African jungle when he was led captive to the last slave ship that sailed from Africa to America.

General Wilson, it is said, wanted to take Frank with him, and Col. Mott let Frank understand that under the Emancipation proclamation he was free to decide for himself, which he promptly did in favor of staying with his former owner, Col. Mott ... Col. Lamar is buried in Laurel Grove Cemetery, Savannah, Ga. The inscription on his monument reads: "Chas. A. L. Lamar. Born in Savannah, April 1st, 1824. Killed during the fight at Columbus, April 16th, 1865." His wife, Caroline Agnes, survived him until 1902.
(Charles Swift paper)

Although the Battle of Columbus ended by 11:00 pm on the evening of April 16th, the destruction of the city was yet to begin. With the Mott House as his headquarters, General Wilson directed General Winslow to destroy and burn all manufacturing facilities in both Columbus and Girard. While these orders were being carried out, Wilson authored an official correspondence concerning his latest victory in battle. Wilson states:

> *My forces captured this place by a most gallant attack 10 o'clock last night, losing 25 men killed and wounded, and captured about 1,500 prisoners, 24 field guns, and 1 gun-boat carrying six 7-inch rifled pieces. General Cobb and 600 [of his] force escaped in the dark. Major-General Upton and Brigadier-General Winslow deserve highest commendation for their personal intrepidity and good management.*
>
> *General Winslow is burning navy-yard, foundries, arsenals, factories, armory, railroad stock, depots, and cotton warehouses today. The value in Confederate currency of property destroyed cannot be estimated. Part of my corps is now moving eastward, and everything will follow in the morning. I anticipate no great difficulty. My command is in magnificent condition. I have just received dispatch from La Grange, commanding Second Brigade, McCook's division. He captured West Point, and fortifications defending it, by assault at 2 p. in. yesterday; killed rebel General Tyler, took 200 prisoners, all the guns, 15 engines, 200 cars, and large quantities supplies.* (OR Series 1, Vol. 49, Part II, p.383)

With Union victories at both West Point and Columbus, General Wilson then focused on Macon and another engagement with General Cobb. Yet, his work in Columbus and Girard was not yet finished. While Colonel Noble served as the Provost Marshall and attended the numerous Confederate prisoners, General Winslow's men engaged in the destruction of anything valuable to the Confederacy.

General Winslow's report of the property destroyed provides valuable details concerning the importance of the Columbus/ Girard facilities and resources, which the area contributed to the Confederacy. His report lists the following data:

Fountain Warehouse: Six thousand bales C.S.A. cotton.
Alabama Warehouse: Seven thousand bales C.S.A. cotton, 100 boxes of tobacco, 20 hogsheads and 100 barrels of sugar, and other commissary stores.

Near Macon Railroad depot: Three large warehouse containing 20,000 sacks of corn, an immense amount of quartermaster's property, commissary stores, and valuable machinery, all in readiness for shipment. A large number of caissons and limbers, generally unserviceable; 100 bales of cotton; also 13 locomotives, 10 passenger cars, 45 box, 24 flat, and 9 coal cars; 1 round-house and machine-shop.

Naval Armory: One small rolling-mill in operation- 1 engine, 40-horse-power; 1 blast engine, 8-horsepower; 2 sets of rollers, and 3 furnaces, capable of making 4,000 pounds of iron per day. One new rolling-mill nearly completed-one 150-horsepower engine, intended to roll railroad and boilerplate iron; 3 large furnaces; 1 blast engine, 10-horsepower; one 10-horsepower steam-hammer.

This building was 150 feet square. One machine-shop-2 engines, 45-inch cylinder, nearly completed, 160 feet shafting; 3 small and 2 large planers; 16 iron lathes; 1 large lathe; 7 feet face plate; 3 drill-presses; 30 vises; 15,000 pounds of brass. All lathes and planers had full sets of tools. One blacksmith shop, containing 10 forges. Several officers and drawing-rooms, with their contents. One pattern-shop, with 3 wood turning lathes and 1 wood-planer. Foundry, boiler-shop, copper-shop, and their contents.

Navy-Yard: Containing brass foundry, boat-building house, and 1 machine-shop, with hot-air furnace; 1 engine, 8-horsepower; 1 large planer; 1 rip-saw and drill-press; 5,000 rounds of large ammunition; also 1 blacksmith shop and tools. McElhaney & Porter's foundry; Containing 1 engine, 20-horsepower.

Niter-Works: Two hundred hands were here employed. Muscogee Iron-Works: Consisting of foundry, machine-shop, small-arms manufactory, blacksmith shop (300 forges), a large saddler's shop, with tools, and 100 sets of flasks; one engine, 30-horsepower.

C.S. Arsenal: Consisting of machine shops, foundries, with two 30-horsepower engines, 2 furnaces a large amount of machinery and war material; blacksmith shop (16 forges).
Two powder magazines: Thirteen thousand pounds of powder, 4,000 loaded shells, 81,000 rounds ammunition for small-arms, and large quantities of rockets, fuses, &c.

Eagle Oilcloth Factory: Four-story brick, 150 feet by 50 feet; 136 looms, 3,450 spindles, cotton, and 1,200 spindles, wool; 2,200 yards of jeans, and 1,500 yards osnaburg made each day.
Howard Oilcloth Factory: Five-story brick building with basement, 120 feet by 50 feet; 146 looms, 5,200 spindles, cotton. This factory made 5,000 yards cloth per day.

Grant Oilcloth Factory: Three stories and basement, brick building, 70 feet by 40 feet; 60 looms and 2,000 spindles, cotton. Made 2,000 yards cloth each day.

Haiman's Iron Foundry: One small engine.

Rock Island Paper Mill: Manufactured printing, letter, and wrapping paper.

Columbus Iron-Works: Sabers, bayonets, and trace-chains were here made; 1,000 stand of arms found. Haiman's Pistol Factory: This establishment repaired small arms, made locks, and was about ready to commence making revolvers similar to Colt army.

Hughes, Daniel & Co.'s Warehouse: Ten thousand bales of cotton. Press and type of following-named newspapers: *Columbus Sun, Columbus Enquirer, Columbus Times*, and the type, one press, &c. of *Memphis Appeal*.

The following is a list of pieces and caliber of artillery which was either partially or wholly destroyed, viz: One 10-inch Columbiad, four 10-pounder Parrotts, one 10-pounder smooth-bore, and eighteen 6-pounder and 12-pounder guns and howitzers, with limbers and caissons (except the Columbiad), all used in the action of the 16th instant and taken while in position.

At the navy-yard were two 6-inch siege guns, mounted, one 30-pounder Parrott, and 4 boat howitzers (brass), not mounted. At the depot were 2 rifled siege guns and 1 smooth-bore siege gun, not mounted: also 11 old iron guns (field pieces), and 2 mountain howitzers, mounted.

Near headquarters post were 4 brass 6-pounders and limbers, smooth-bore, and at a foundry northeast part of town were 16 field pieces, caissons, &c., caliber not known. At the arsenal was 1 Napoleon gun, new, quite a number of limbers and caissons. Total number of guns, exclusive of the 6 splendid 7-inch rifled ones on gunboat Jackson, 68. Nearly all were thrown into the river.

Quartermaster's property found in store and issued to the troops and Negroes or destroyed: 4,500 suits of Confederate uniform, 5,890 yards army jeans, 1,000 yards osnaburg, 8,820 pairs of shoes, 4,750 pairs of cotton drawers, 1,700 gray jackets, 4,700 pairs of pants, 2,000 pairs of socks, 4,000 tin cups, 2,000 tin plates, 960 wooden buckets, 20 telegraphic instruments, 400 shirts, 375 hatchets, 650 gray caps, 33 tin pans, 6 coils 1/2-inch rope, 15 boxes carpenter's tools, 400 wall-tents and flies, 1,000 axes and halves, 1,000 picks and halves, 400 spades and shovels.

Destroyed at Girard (opposite Columbus): One rope factory, 2 Government blacksmith shops, 2 locomotives, 15 boxcars, and an extensive roundhouse and railroad machine shop. The machine shops, foundries, factories, and other works destroyed here, as above enumerated, and were of immense value to the rebels and to the entire South.

More than 5,000 employees are thrown upon the community for other support. No private buildings are thrown upon the community for other support. No private buildings in Columbus were destroyed, and no buildings fired except by order and with proper authority… Four bridges over the Chattahoochee River, at and near Columbus, were thoroughly destroyed, one (old) by the enemy and three (including the railroad bridge) by our troops. (Series 1, Vol. 49, Part 1, pp. 486-87, Winslow's report)

While no private buildings were intentionally destroyed, the huge fires and explosions of munitions burned and roared through the community for days. Stores and homes were plundered for anything of value by both Union troops and local citizens. This vast destruction of property was devastating to the region's economy, and many citizens faced immediate, long-term poverty. According to Harold Coulter, the citizens of Girard suffered worse fates than the people in Columbus, as Girard served as the main battlefield. Coulter explains with:

There was little left on the Alabama side. All the businesses and factories in Girard were destroyed as were several of the more substantial houses and all that were left was a cluster of houses in Holland Creek about two blocks west of the river, another cluster, these of shoddy and disreputable construction sitting in the midst of a red-light district and "houses of disrepute," located about a half mile below Holland Creek on a hill overlooking the river.

There were two substantial houses and three tenant houses known to exist at the time on Moses Hill (Red Hill) Newspaper accounts mention "scores of empty houses, shacks, and shanties in Girard." The result of scores of families leaving the area for safety and food in other locations. Also, many of them moved across the river into Columbus to be eligible for the food and other supplies (government rations), which were doled out every two weeks. (Coulter)

Both Columbus and Girard suffered economically for some time to come but the immediate concern was food, which Wilson's men took freely from homes. They also destroyed vast quantities of corn and with the loss of the factories; there would be no employment for a long time. This was all especially frightening and stressful for families with mouths to feed.

One of Wilson's men recorded that, "the women and children who had been employed in the factories and arsenals turned out with one accord, to pillage the stores and the Government warehouses. (OR Series 1, Vol. 49, Part I, pp. 407-409 Salter's Report)

General Wilson also noted the situation with, "There are thousands of almost pauper citizens and Negroes, whose rapacity under the circumstances of our occupation, and in consequence of such extensive destruction of property, and was seemingly insatiable. The citizens and Negroes formed one vast mob, which seized upon and carried off almost everything movable, whether useful or not." (Series 1, Vol. 49, Part 1, pp. 486-87, Winslow's report)

The women were certainly concerned for the well-being of themselves and their families, as their men were, dead, wounded, or imprisoned. The Yankees were burning the town, and desperately, the women of the community struggled to obtain what remnants of food and clothing that could be had. One eyewitness that observed the looting in downtown Columbus wrote, "There is evident demoralization among the females. They frantically join and jostle in the chaos, and seem crazy for plunder. There are well-dressed ladies in the throng." (Mitchell, p.192)

Women also hid what valuables they owned at home, lest the Union soldiers, or local criminals steal them. According to one lady:

The women and children had anxiously watched during the battle over in Girard, which looked like a sheet of flames along the Chattahoochee... Hastily they had concealed jewelry and valuables, and buried their silver. They put it under mattresses, sent some of it to plantations, and some silver, as in the case of the Redd's [family of Albert Redd], was buried in Linwood Cemetery." (Worsley, p. 297)

Many stories abound concerning the "evil Yankees" and their wanton destruction delivered upon Columbus, but Wilson's men were well disciplined, and the destruction of Confederate property was a legal and valid order. While isolated incidents of a dubious nature occur during war, any acts committed in Columbus pale in comparison to previous raids such as Sherman's infamous "march to the sea."

Moreover, in addition to the destruction of property in Columbus and Girard, the Union soldiers, searched homes, confiscated food, and impressed horses, mules, and carriages.

Yet, these acts were carried out under the authority of martial law and within wartime constraints. In fact, there were no churches, schools, homes, or private properties destroyed intentionally by Wilson's men. Moreover, while the downtown section of Columbus was indeed vandalized, the citizens of Columbus perpetrated this action.

There are several diaries of Union soldiers that indicate isolated incidents of illegal conduct wherein individuals stole private property, and a few soldiers participated in the pillage of downtown Columbus. Yet, General Wilson was a strict disciplinarian and he punished the perpetrators accordingly. (Hinricks Diary & Harrington Diary, April 28).

In 1864, the federal government established a system to reimburse citizens for property losses incurred during the war; however, the government did not provide relief in the southern states until 1871. The Southern Claims Commission operated from 1871 until 1880 investigating and adjudicating claims for compensation due to destruction or forfeiture of property at the hands of the US government.

Ultimately, the commission approved a total of seven claims submitted from citizens of Muscogee County, Georgia, and most of these involved impressments of horses, mules, and carriages. In one such case, a free black named Joe Clark surrendered 5 mules, 2 buggies, and 2 wagons to federal troops.

On the 17th of April, Mr. Clark's family and friends rode into town to view the unfolding events, and when they reached the train depot, a Yankee officer took their property and told Mr. Clark if he "did not leave there, he would take the top of his head off!"

Interestingly, one of the claimants was Dr. Thomas S.

Tuggle, the suspected member of the local "Peace Society," whose claim for reimbursement of 1,000 bushels of burnt corn was promptly approved. (National Archives, Record Group 217, Microfiche # M1658, Fiche # 1, 2, & 3; Claim Number 15615)

Throughout the evening of April 17th, Wilson's men continued their duties of destroying Confederate property, as Columbus was the second largest storehouse of the Confederacy. The Union soldiers were amazed at the wealth of resources in Columbus, as they had believed Selma was a larger depot, but they were mistaken. Captain Hinricks recorded, "The shells were exploding the whole night and every now and then a magazine was blown up and made the earth tremble for miles." (Hinricks, April 17)

Another soldier, Sergeant Harrington wrote in his diary on the same night that, "Tonight the city is brilliantly illuminated by the fire of the public buildings, which with the noise of bursting shells makes a grand scene." (Harrington, April 17) According to General Wilson, "The destruction of the last factories, depots, and warehouses of the Confederacy was as complete as fire could make it, and of itself must have been the deathblow to the Confederacy, even if it had been able to keep its armed forces together for a further struggle. (Wilson, p. 268)

That night, the flames that consumed the warehouses and factories in the Chattahoochee Valley marked the end of the war. After Columbus and Girard were torched, there was no more wholesale destruction of Confederate property, and in hindsight, the Union commanders reflected that if they had only known of events in the east, it would not have happened.

On April 18 at 8 o'clock, Wilson's Cavalry Corps continued their eastward trek with a new objective, Macon, Georgia. (Hinricks Diary, April 18; OR Series 1, Vol. 49, Part 1, p. 407-09, Report of Surgeon F. Salter) That morning Captain Hinricks noted that, "As we passed through town, it was a sight to behold, the whole city on fire- the people carrying their plunder from place to place- the shells still exploding. We struck the Macon Road and then we knew that Macon was our next point- another hard nut- I hope not too hard." (Hinricks Diary, April 18)

With their Confederate prisoners in tow, Wilson's Corps moved towards Macon, Georgia with the same resolve they had attacked previous Confederate strongholds, yet somehow, the participants knew the end was on the horizon.

Before leaving, General Wilson placed the ranking Confederate, Colonel James C. Cole in charge of Columbus. Messages housed in the Official Records of the war between Colonel Cole and General Wilson reflect the two professional soldiers had reached an honorable, gentlemen's agreement concerning the situation. Colonel Cole carried out his final duties of the war distributing rations and organizing safety until he was finally replaced in mid- May by Union officers.

General Wilson had to continue his mission, which necessitated moving his entire corps to Macon. Colonel Cole's actions display a sincere desire to maintain order. However, his duties were thankless tasks amidst an embittered people and his authority was immediately challenged.

Once the Union troops withdrew from Columbus a detachment of Buford's cavalry returned to Columbus. Seeing that Mr. Randolph Mott's cotton warehouse was the only one that General Wilson had not burned, the Confederate cavalrymen torched it. (Wilson, Telfair) While this act of retribution was unwarranted, anger and desperation filled the chaotic void of what once was the Confederacy. In defeat, the chivalry and honor that once represented a proud people was replaced with vengeance and cruelty.

While Buford's men were in Columbus "assisting the

commandant of the post [Colonel James Cole] in restoring order," an unnamed local citizen tried to steal a horse belonging to one of the cavalry officers. Captain Addison Harvey, commander of "Harvey's Scouts" embarrassed the perpetrator, and then went about his business.

In an act of reprehensible violence, the horse thief snuck up behind Captain Harvey and shot him in the back of his head. Captain Harvey was young, well respected, and a dashing cavalry officer that had survived the war, unwounded.

Tragically, although the war was over, he was the last Confederate to die in Columbus. Killed by a fellow southerner, Captain Harvey was buried in Linwood Cemetery. (Columbus Sexton's Reports, Book E, 1853-1866, City Council records, City Clerk's Office; OR, Series 1, Vol. 49, Part 2, P. 1271-72).

General Buford and his few remaining Cavalry troops departed Columbus and headed towards western Alabama in hopes of rejoining General Forrest.

When General Wilson drove towards Macon, he remained ignorant concerning the capitulation of the Confederacy and on April 20th, his men faced their final battle despite "rumors" of peace. After skirmishing with remnants of the Georgia State Line, 15 miles west of Macon, the Union column was met by a flag of truce borne by General Felix Robertson, one of Wilson's West Point classmates.

While Robertson carried news that Generals Sherman and Johnston were negotiating terms of surrender in North Carolina, Wilson's lead commander told him, "I'll give you just five minutes to get out of the way with your flag of truce and escort." (Wilson, p. 277)

The message was sent back to Division and Corps headquarters, but the Union column continued their mission to take Macon. General Wilson rode forward when he received the news, and he arrived in Macon where he met with General Howell Cobb.

The Confederate troops in Macon surrendered to the Union troops, but under a violent protest as they expected better treatment. The confusion was simply a matter of miscommunications. Generals Cobb and Wilson sorted out the details by the next day after receiving clarification via telegraph from General Sherman. (Wilson, p. 285-286)

On May 10th, a detachment of Wilson's cavalry apprehended the former Confederate president, Jefferson Davis and his wife at Irwinville, Georgia. Finally, the Civil War was officially over in Georgia, and except for the parole of the remaining Confederate troops, the men would soon return to their former occupations, and begin the task of rebuilding their nation.

In Columbus, "Hundreds of idle workmen walked the streets, their number augmented by freed Negroes who had nothing to do, and made their living by looting and stealing from the whites." (Telfair, p. 144)

It was impossible for Colonel Cole in "command" of a defeated army to suppress the violence. In a message to a nearby Union officer, he pleaded for assistance with:

General Wilson promised me that if my forces were not sufficient to suppress marauding parties in my vicinity during the present armistice, to furnish me a force to cooperate with him for that purpose. Can you do the same immediately, until I can get a courier to him? A strong force is necessary at this post to suppress riot and preserve public stores as well as private. (OR, Series 1, Part 2, p. 597)

General James Wilson and his staff officers.

In other correspondence with General Wilson, Colonel Cole arranged for distribution of food along with other humanitarian, logistical, and administrative matters until General Wilson could send a detachment of Union troops back to Columbus.

It was not until May 17th that Captain John C. Lamson of the Seventeenth Indiana (mounted) Infantry arrived in Columbus and relieved Colonel Cole of his arduous duties.

Over the next few weeks, Captain Lamson paroled 3,700 Confederate soldiers in Columbus, and eventually the region returned to normal. By July of 1865, several business and factories were back in business, and like the mythical Phoenix, Columbus and Girard also "rose from the ashes."

Meanwhile, General Forrest had regrouped the bulk of his command in the rural community of Gainesville, Alabama. This tiny hamlet lies in the western portion of Alabama, on the western shore of the Tombigbee River, near the border with Mississippi. While the men were ready to follow Forrest into battle once more, rumors and uncertainties created an anxious situation.

While encamped at Gainesville, newspapers confirmed rumors regarding the fall of the Confederate government, the assassination of President Lincoln, the surrender of Generals Lee and Johnston's Armies, the fall of Richmond, Mobile, Montgomery, and the huge success of Wilson's raid through Alabama and Georgia.

In fact, Lieutenant General Taylor had already surrendered his Confederate Department to Union Lieutenant General Edward Canby, so, for Forrest and his men, the war had already ended, they just didn't know it.

On May 9, 1865, Brigadier General Elias S. Dennis arrived in Gainesville as the official federal commissioner to issue paroles to the demoralized Confederate Cavalry.

General Forrest relinquished his authority and he addressed his troops with these final words:

Soldiers: By an agreement made between Lieutenant-General Taylor, commanding the Department of Alabama, Mississippi, and East- Louisiana, and Major-General Canby, commanding United States forces, the troops of this Department have been surrendered.

I do not think it proper or necessary, at this time, to refer to the causes which have reduced us to this extremity"; nor is it now a matter of material consequence to us how such results were brought about that we are beaten, is a self-evident fact, and any further resistance on our part would be justly regarded as the very height of folly and rashness.

The armies of Generals Lee and Johnston having surrendered, you are the last of all the troops of the Confederate States Army, east of the Mississippi River, to lay down your arms.

The cause for which you have so long and so manfully struggled, and for which you have braved dangers, endured privations and sufferings, and made so many sacrifices, is to-day hopeless. The Government which we sought to establish and perpetuate is at an end. Reason dictates and humanity demands that no more blood be shed.

Fully realizing and feeling that such is the case, it is your duty and mine to lay down our arms—submit to the 'powers that be'—and to aid in restoring peace and establishing law and order throughout the land.

The terms upon which you were surrendered are favorable, and should be satisfactory and acceptable to all. They manifest a spirit of magnanimity and liberality on the part of the Federal authorities, which should be met, on our part, by a faithful compliance with all the stipulations and conditions therein expressed.

As your Commander, I sincerely hope that every officer and soldier of my command will cheerfully obey the orders given, and carry out in good faith all the terms of the cartel.

Those who neglect the terms, and refuse to be paroled, may assuredly expect, when arrested, to be sent North and imprisoned. Let those who are absent from their commands, from whatever cause, report at once to this place, or to Jackson, Mississippi; or, if too remote from either, to the nearest United States post or garrison, for parole.

Civil war, such as you have just passed through, naturally engenders feelings of animosity, hatred, and revenge. It is our duty to divest ourselves of all such feelings; and, as far as in our power to do so, to cultivate friendly feelings toward those with whom we have so long contended, and heretofore so widely, but honestly, differed.

Neighborhood feuds, personal animosities, and private differences should be blotted out; and, when you return home, a manly, straightforward course of conduct will secure the respect even of your enemies. Whatever your responsibilities may be to government, to society, or to individuals, meet them like men.

The attempt made to establish a separate and independent Confederation has failed; but the consciousness of having done your duty faithfully, and to the end, will, in some measure, repay for the hardships you have undergone.

In bidding you farewell, rest assured that you carry with you my best wishes for your future welfare and happiness. Without, in any way, referring to the merits of the cause in which we have been engaged, your courage and determination, as exhibited on many hard-fought fields, have elicited the respect and admiration of friend and foe. And I now, cheerfully and gratefully, acknowledge my indebtedness to the officers and men of my command, whose zeal, fidelity, and unflinching bravery have been the great source of my past success in arms.

I have never, on the field of battle, sent you where I was unwilling to go myself; nor would I now advise you to a course which I felt myself unwilling to pursue. You have been good soldiers; you can be good citizens. Obey the laws, preserve your honor, and the Government to which you have surrendered can afford to be, and will be, magnanimous.

Buford's Alabama cavalrymen arrived at Gainesville on May 14th, and by May 16th, the final surviving elements of Forrest's Cavalry had laid down their arms, and returned home.

As a civilian, Nathan Bedford Forrest served as a railroad executive on several southern lines. Known as one of the finest cavalry officers in history, southerners still proudly boast of their ancestors who "rode with Forrest." He passed away on October 29, 1877 but his legacy is alive and well.

By the end of the summer of 1865, the bulk of Wilson's Cavalry Corps were released from active federal service and they returned to their homes in the north.

Their commander, General James H. Wilson continued his career as an Army officer during the Chinese Boxer Rebellion and on into the Spanish American War. He also helped connect the nation back together as a railroad executive. He died on February 23, 1925 and he is buried in Wilmington, Delaware.

Table 4

Casualty Lists (Union and Confederate):
Recorded in the CSA Hospital Logs at
Columbus, Georgia (April 16-17, 1865)

NAME	UNIT	Comments
Confederate Casualties		
Lieutenant Joseph **Ashman**	25th Georgia Infantry	
Private John **Arnett**	3rd Alabama Reserves	Gunshot
Sergeant James **Arnold**	20th Alabama Infantry	Gunshot
Private L. **Babarne**	Alabama Reserves	
Sergeant Frances **Baker**	Alabama Reserves	Gunshot
Lieutenant D.E. **Banks**	2d Georgia State Line	Gunshot
Private Milton **Barnes**	57th Alabama Infantry	
Private William **Barally**	3d Alabama Reserves	
Private John **Barry**	Cole's Battalion	Local Alabama reserves served under Captain Cole and COL Bratton
Private J.G. **Bice**	Cole's Battalion	
Private Jenkins **Benton**	Alabama Reserves	
Lieutenant J.H. **Birch**	Waddell's Battery	Alabama 20th Artillery Battalion- "Waddell's"

203

Private Julius **Brieson**	Cole's Battalion	
Private Miles **Bloodsworth**	15th Alabama Cavalry	
Private Joel **Bleason**	3d Georgia Reserves	Gunshot
Private Thomas **Blackshear**		
Private William **Borally**	3d Alabama Reserves	
Lieutenant John **Bailey**	2nd Georgia State Line	Gunshot
Private Augustus **Beakles**		
Private M.D. **Bostick**	Cole's Battalion	
Private W.J. **Brisbee**	2d Georgia Cavalry	Gunshot
Private Marlie **Bustie**	Cole's Battalion	
Private Joseph **Caldwell**		
Private William **Colquitt**		
Private Charles **Carpenter**	Jacques Battalion	MAJ Samuel R. Jacques Battalion, 1st Columbus City Infantry Battalion
Private J.C. **Chilas**	Waddell's Battery	
Private James **Cook**	Waddell's Battery	
Corporal James T. **Cook**	3d Georgia Cavalry	
Private Jesse **Cone**	2nd Alabama Reserves	

Private John **Carrington**	3rd Alabama Reserves	
Private James E. **Cargill**		
Private S. M. **Crew**	Arsenal Battalion	Columbus militia unit
Private William **Camp**	3d Alabama Reserves	
Private H. G. **Chapman**	Jacques Battalion	
Private Robert **Castin**	Local Alabama reserves served under Captain Cole and COL Bratton	
Private A. **Cobb**		
Private John C. **Calhoun**	Jacques Battalion	
Private James E. **Cargill**	1st and 2nd Georgia State Line comprised the Georgia State Militia	
Private W.J. **Crutcher**	27th Alabama Infantry	
Corporal Hamilton **Dewberry**	46th Georgia Infantry	Gunshot
Private James L. **Dickenson**	4th Alabama Infantry	
Private James **Down**	Cole's Battalion	
Private G.W. **Davis**	Clanton's Battery	
Private William H. **Day**	Waddell's Battery	
Private Andrew J. **Dowdy**		
Private Frederick **Decker**	Bates escort	Tennessee Cavalry Unit
Private John Dowd	8th Alabama Cavalry	

Private George **Driver**	10th Texas Cavalry	Gunshot
Private William J. **Ellis**	3d Alabama Reserves	
Private William C. **Ellis**	3d Alabama Reserves	
Private Lewis **Estigent**	4th Alabama Reserves	
Captain Levi **Godwin**	3d Alabama Reserves	
Private John **Gallager**	Cole's Battalion	
Private J. **Greenhow**	Cole's Battalion	
Private D. L. **Green**	Waddell's Battery	
Private Joseph **Garrett**	7th Alabama Cavalry	
Private Gustavus **Gnoppelous**		
Private John **Greene**	Cole's Battalion	
Private John **Gaffney**	Waddell's Battery	
Captain John **Gamble**	3rd Alabama Reserves	(Gunshot- back)
Captain John S. **Pemberton**	Pemberton's Cavalry	(Gunshot-chest, saber wound also)
Corporal Samuel **Garlick**	Waddell's Battery	
Private John **Gehrit**	7th Alabama Cavalry	
Private A. J. **Grant**	3d Alabama Reserves	
Private William **Hair**	31 Mississippi Infantry	

Lieutenant William **Hall**	9th Georgia Battalion Artillery	
Private William **Hare**	31st Mississippi	
Sergeant William **Honeyfin**	Waddell's Battery	Gunshot
Private Wiley **Houston**	37th Mississippi	
Private George **Hamilton**	29th Tennessee Infantry	(Gunshot-head)
Private Ned **Hardy**	3d Conf.	Gunshot
Private William **Harrell**	4th Alabama Reserves	
Private J.K. **Harris**	Waddell's Battery	
Private W.H. **Holland**	Conscript	
Captain Charles **Hooper**	Clanton's Battery	
Private James **Howard**	9th Ark Infantry	
Private Lucian **Haney**	4th Alabama Reserves	
Private S.H. **Hutchins**	26th Georgia Battalion	(Company B)
Private George **Ingraham**	26th Georgia Battalion	(Company B)
Private George **Ivey**	2nd Georgia State Line	Gunshot
Private E.J. **Kiplinger**	Waddell's Battery	
Sergeant Arbell **Koll**	3d Alabama Reserves	
Private Moses **Lee**	Alabama Reserves	Gunshot

Lieutenant E.D. **Lee**	Alabama Reserves	
Private William **Leonard**	Waddell's Battery	
Private Jesse **Logan**	Alabama Reserves	
Private James **Lowe**	Alabama Reserves	
Lieutenant Edward **Lee**	Alabama Reserves	
Private Thomas **Miller**	3d Georgia Cavalry	Gunshot
Private William **Minkious**	46th Georgia Infantry	
Private Christopher **Morgan**	3d Alabama Reserves	
Private Littleboy **Matthews**	3d Alabama Reserves	
Private M.J. **Moore**	18th Alabama Infantry	Gunshot
Lieutenant J.C. **Moore**	Alabama Reserves	(Died April 20, 1865)
Private Dennis **McLauri**		
Private John **Moore**	Alabama Reserves	(Gunshot- head)
Private John **Minor**		
Private Matthew **McNeil**	5th Conf.	
Private George W. **Morgan**	9th Georgia Battalion Artillery	Gunshot
Private John **Morgan**	3d Alabama Cavalry	
Private Larry **Moore**	5th Conf.	

Private Newell **Moore**	22nd Alabama Infantry	
Private Charles **McClendon**	22nd Alabama Infantry	(Gunshot- head)
Private Jesse **McCall**	Waddell's Battery	
Private E.L. **Mullins**	44th Mississippi Infantry	Gunshot
Sergeant E.M. **Murphy**	24th Texas Infantry	Gunshot, (Company D)
Private William **Nance**	2nd Georgia Cavalry	
Private James **Newman**	12th Georgia Cavalry	Gunshot
Private Jerry **Newman**	5th Conf.	Gunshot
Private Benton **Peters**	15th Alabama Infantry	
Private E.W. **Peters**	15th Alabama Infantry	
Private William **Parker**	6th Alabama Reserves	
Private Charles **Pritchitt**	3d Alabama Reserves	(Gunshot- right eye)
Private Richard **Rainey**	12th Georgia Battalion Artillery	(Gunshot, Died July 30, 1865 from wounds)
Private William **Rumbo**	5th Alabama Reserves	(Gunshot- hip)
Private Thompson **Rodgers**	5th Alabama Reserves	
Private Frances **Rumbo**	15th Alabama Infantry	
Private Amos **Ryan**	2d Alabama Cavalry	(Gunshot- knee)
Private Richard **Spencer**	37th Georgia Infantry	

Private Jonathan **Sanders**	3d Alabama Reserves	
Private Richard **Sevant**	4th Alabama Reserves	
Private B.J. **Smith**	3d Georgia Reserves	
Private Edward **Smith**	10th Texas Cavalry	(Gunshot- left side)
Private Calvin **Stewart**	12th Louisiana	Gunshot
Private John **Thornton**	17th Alabama Infantry	
Private D.E. **Thrash**	26th Georgia Battalion	(Gunshot, Company B)
Private Asa **Tyce**	2d Georgia State Line	
Private Benjamin **Watson**	14th Mississippi	
Private James S. **Webb**	5th Conf.	
Private James **Welch**	6th Alabama Cavalry	
Private John **Williams**	15th Alabama Infantry	
Private Michael **Williams**	2d Georgia State Line	
Private E. **Williamson**	Jacques Battalion	
Private John W. **Willhelm**	Waddell's Battery	
Private James **Wilson**	5th Conf.	(5th Confederate was the 40th Tennessee Infantry Regiment)
Private John **Wilson**	3d Alabama Reserves	

Private John **Wilson**	Cole's Battalion	
Private William **Wilson**	5th Alabama Reserves	
Private Edward **Woolfolk**	3d Georgia Cavalry	
Sergeant John **Wilkes**	5th Georgia Reserves	Gunshot
Private John J. **Wallace**	23d Mississippi Infantry	Gunshot
Private William **Young**	26th Georgia Battalion	(Company B)
Colonel Charles **Lamar**	(KIA- upper bridge, Columbus, Georgia, buried in Savannah)	
Private Alexander W. **Robinson**	(KIA- upper bridge, Columbus, Georgia buried in Linwood Cemetery)	
George Hooper **Waddell**	Alabama Reserves	(KIA- Russell County Judge & brother-in-law of COL James F. Waddell, KIA on Broad Street)
J.J. **Jones**	(KIA- Editor of the Enquirer, KIA on Broad Street)	
Captain S. Isidore **Guillet**	(KIA- Girard, Alabama (Von Zinkin's Adjutant- buried in Linwood Cemetery)	
Private Henry A. **Nuckolls**	2d Georgia State Line	(KIA- Girard, Alabama, buried in Forsythe County, Georgia)
Evan **Jones**		(KIA- probably CS Navy- from Apalachicola, FL, KIA on Broad Street)

William **Smith**	(KIA- He and two other men were "killed in the brickyards")
"George"	William Crook's servant (Gunshot-leg)

NAME	UNIT	Comments
Union Casualties		
Private Nathan **Beezley**		KIA- Apr. 16, 1865, Columbus, Georgia, Company I, (buried at Andersonville National Cemetery)
Corporal Daniel **Beers**	3d Iowa Cavalry	Amp. Left leg on Apr. 19th, died in Macon, July 14, 1865, (buried at Andersonville National Cemetery)
Private M.J. **Benge**	4th US Artillery	Gunshot- left leg
Private Jacob **Cellan**	3d Iowa Cavalry	Amp. Left leg, died in Macon on June 26, 1865, Company A, (buried at Andersonville National Cemetery)
Private Charles **Center**	3d Iowa Cavalry	Company A
Private Edwin **Culver**	3d Iowa Cavalry	Gunshot, Quartermaster
Sergeant John Wesley **Delay**	3d Iowa Cavalry	KIA- Apr. 16, 1865, Columbus, Georgia, Company I, (buried at Andersonville National Cemetery)
Private W.H. **Denoya**		Company B, 5th Iowa Cav, died of wounds and buried at Andersonville National Cemetery
Private Benjamin **Grant**	3d Iowa Cavalry	Company B, survived the war
Private Charles **Hartwell**	3d Iowa Cavalry	

Private Thomas **Heirno**	5th Iowa Cavalry	Gunshot- Left leg
Private A. **Hynott**	3d Iowa Cavalry	
Sergeant Joseph H. **Jones**		KIA- Apr. 16, 1865, Columbus, Georgia, Company L, (buried at Andersonville National Cemetery)
Private Edward **Moore**	3d Iowa Cavalry	Company A, survived the war
Private James M. **Miller**	3d Iowa Cavalry	Gunshot- Pelvis, died Apr. 20th, Company K, (buried at Andersonville National Cemetery)
Captain Thomas D. **Miller**	3d Iowa Cavalry	KIA- Apr. 16, 1865, Columbus, Georgia, Company D, (buried at Andersonville National Cemetery)
Private V. **Nuff** (Bugler)	4th US Artillery	KIA- April 16, 1865, Columbus, Georgia, Company D (buried at Andersonville National Cemetery)
Private John **McFauland**	1st Ohio Cavalry	
Private John **McGrath**	4th US Artillery	Gunshot- Left leg
Private Joseph **Martin**		
Private Charles **Nevel**	7th Ohio Cavalry	
Private Samuel **Nelson**	3d Iowa Cavalry	KIA- Apr. 16, 1865, Columbus,

		Georgia, Company I, (buried at Andersonville National Cemetery)
Captain Daniel **Beavers**	3d Iowa Cavalry	
Corporal Henry **Langsford**	3d Iowa Cavalry	Gunshot- Left arm
Private Miles **King**	3d Iowa Cavalry	Gunshot- Abdomen, died Apr. 20th, Company B
Private Richard **Porter**	Co. G, 5th Iowa Cav	KIA- April 16, 1865
Private John **Randolph**	3d Iowa Cavalry	Gunshot- Shoulder, survived the war
Private Samuel **Ritter**	3d Iowa Cavalry	
Private Nathan **Rolland**	1st Ohio Cavalry	Gunshot- left (?)
Sergeant John **Ritchey**	10th Missouri Cavalry	Gunshot- leg
Private John **Shirley**	4th Iowa Cavalry	Gunshot- Abdomen (left side)- Company A
Private Pat **Stanton**	4th Iowa Cavalry	Gunshot- left arm
Private John **Stephens**	3d Iowa Cavalry	Gunshot- left leg - Company D
Private William **Sullivan**	3d Iowa Cavalry	Survived the war
Corporal David **Thompson**	3d Iowa Cavalry	Company C
Private Abram **Wishard**	3d Iowa Cavalry	Gunshot- right side of face, survived the war

Private Abram **Webster**	3d Iowa Cavalry	
Private Lewis **Walter**	4th US Artillery	Contusion
Private William **Warwell**	4th Iowa Cavalry	Gunshot- right hand
Private James S. **Wilson**	3d Iowa Cavalry	Company M, survived the war
W. **Austin**	3d Iowa Cavalry	Company K, POW at Andersonville, died Aug. 8, 1864
D.S. **Beers** (Daniel)	3d Iowa Cavalry	Company D, WIA- Columbus, Georgia, died July 14, 1865- Macon, Georgia
N. **Beezly** (Nathan)	Company I, 4th Iowa Cav	KIA at Columbus, Georgia, Apr. 16, 1865
J.K. **Billings**	5th Iowa Cavalry	Company B
J. **Cellan** (Jacob)	3d Iowa Cavalry	Company A, WIA- Columbus, Georgia, died June 26, 1865- Macon, Georgia
W.A. **Cox**	5th Iowa Cavalry	Company I
H. **Cox**	5th Iowa Cavalry	Company I
J.W. **Delay** (John)	3d Iowa Cavalry	Company I, KIA at Columbus, Georgia, April 16, 1865

W.H. **Denoya**	Company M, WIA- Columbus, Georgia, died ___?___ - Macon, Georgia	
J.S. **Ireland**	3d Iowa Cavalry	Company I
J.H. **Jones** (Joseph)	Company L, KIA at Columbus, Georgia, April 16, 1865	
L.J. **Littlejohn**	5th Iowa Cavalry	Company B
D.C. **Lauderbeck**	5th Iowa Cavalry	Company B
J.B. **Martin**	5th Iowa Cavalry	Company B
J.A. **Mercer**	4th Iowa Cavalry	Company I
T.J. **Miller** (Thomas)	3d Iowa Cavalry	Company D, KIA at Columbus, Georgia, April 16, 1865
S. **Sutton**	5th Iowa Cavalry	Company I
J.A. **Whiten**	5th Iowa Cavalry	Company I

END NOTES

Abbreviations used in notes:
OR: Official Records of the War of the Rebellion
GC: George Greene Collection

CHAPTER 1
None

CHAPTER 2

1. Battles & Leaders of the Civil War, Vol. 4, p. 465
2. OR 45-1, Report # 194
3. OR 45-2, January 2, 1865. Brig. Gen. A. B. Dyer, Chief of Ordnance
4. Taylor, 218
5. OR 45-2, January 3, 1865
6. Wills, quoted in 301-02; Wilson, p. 184
7. OR 49-1, p.994
8. Harrington Diary (February- April, 1865)
9. Gilpin, 618
10. OR 49-1. p. 310
11. OR 49-2, p. 1160
12. OR 49-1, p. 358-9
13. Taylor, p. 219
14. OR 49-1, p. 360
15. Wilson, Vol. 2, p. 241-43
16. Harrington Diary, April 4-8, 1865
17. Rogers, p. 145-7
18. OR 49-2, p. 1239
19. OR 49-2, p. 405, Wilson, Vol. 2, p. 250-53
20. OR 49-2, p. 344
21. OR, ser. 1, Vol. XLIX, Pt. 1, P. 364; Wilson, Vol. 2, p. 265
22. Harrington Diary (April 1865)

23. Hinricks Diary, April 13, 1865
24. Larson, p. 301
25. Greene Collection
26. Hinricks Diary, April 14, 1865
27. Gilpin, 651
28. Larson, p. 208
29. Hinricks Diary, April 15, 1865
30. Scott, p. 482
31. Yeoman, p. 222-23

CHAPTER 3

1. Muscogee County Statistics (Muscogee County, Georgia-Mayor's Office)
2. Dameron, 89
3. Greene Collection
4. Coulter, p. 116
5. Martin, p. 56-7
6. Greene Collection; Standard, p. 15
7. Martin, p. 94 and 104
8. Columbus (Georgia) Enquirer, April 10, 1860; Martin, Columbus, Georgia, From Its Selection as a "Trading Town" in 1827, To Its Partial Destruction by Wilson's Raid in 1865. Volume 1, p. 118; Daily Sun (Columbus), March 16, 1861; April 8, 1863; Standard, Columbus, Georgia in the Confederacy: The Social and Industrial Life of the Chattahoochee River Port, pp. 28-9, (detailed narrative of Georgia's war time industrial production).
9. Greene Collection
10. Dameron, p 110-15; Mosocco, The Chronological Tracking of the American Civil War per the Official Records of the War of the Rebellion, p. 6 (excellent reference source for accurate dates of events during the Civil War); The Daily Times (Columbus), January 13, 1861; Worsley, p. 272, Columbus on the Chattahoochee; Ibid., February 21, 1861; The Daily Sun (Columbus), April 17, 1861; Worsley, Columbus on the Chattahoochee, p. 272; Columbus (Georgia) Enquirer, January 2, 3, 1861, March 20, 1862; The Daily Sun, January 1 and 31, 1861, April 1, 1861.
11. Dameron 118; Columbus Daily Enquirer, June 20, 1861
12. OR, Series 1, Vol. 23, part I, p. 283-85
13. Confederate Veteran Vol. XXXV, April 1927, p. 138
14. Standard, p. 28-9
15. Ibid., p. 32 and 36
16. Greene Collection
17. Columbus Daily Sun, June 10, 1861
18. Greene Collection

19. Columbus Enquirer, August 12, 1864
20. Standard, p. 38-9
21. Greene Collection
22. Ibid.
23. Columbus Times, November 17, 1861
24. Albaugh, p. 57
25. Greene Collection
26. Columbus Daily Sun April 8, 1863; November 1864; Standard, p. 42

CHAPTER 4

1. OR Series II, Vol. 2, pp. 750-3; Columbus Enquirer November 6, 1864
2. Columbus Enquirer, March 23, 1862; Sun, June 5, 1862; Standard, p. 43
3. Stanford letter; Navy Grey; Columbus State University Archives (Rosters of Confederate Employees- Columbus Naval Yard: 1863-1865)
4. Columbus Enquirer, July 21, 1863
5. Ibid., July 28, 1863
6. OR Series I, Vol. 39, Part 2, p. 882; Vol. 45, Part 2, p. 644
7. Columbus Enquirer, November 10, 1863
8. OR Series I, Vol. 28, part II, p. 554
9. Ibid, p. 581
10. Saltmarsh letter, Mandrell p. 344-50
11. OR, Series I, Vol. 14, Part 1, p. 681- 682; Series I, Vol. 28, part 2, p. 279, Series I, Vol. 52, Part 2, p. 375
12. Mandrell, p. 344
13. Columbus Enquirer April 28, 1864
14. OR Series 1, Vol. 28, part 2, p. 554
15. OR Series 2, Vol. 7, p. 518-19
16. Columbus Enquirer May 20, 1864; June 10, 1864
17. Greene Collection
18. Muscogee County Court Records (Local Troop rosters- Probate Court Records)
19. Bragg, p. x, xi, 128-161
20. OR, Series 4, Vol. 3, p. 55
21. OR, Series 4, Vol. 3, p. 463
22. OR, Series 4, Vol. 3, p. 463
23. Columbus Enquirer, July 30, 1864
24. Ibid., September, 9, 1864
25. Ibid., September 11, 1864
26. Ibid., September 12, 1864
27. Telfair, 133; Worsley, 294
28. Georgia Archives, Pension Records, Widow of John Lindsay; Worsley, p. 293

29. OR Series 4, Vol. 3, p.67
30. OR Series 4, Vol. 3, 393-96; Series 1, Vol. 39, Part 2, p. 585-90
31. Columbus Enquirer Bill Winn; Scott, Upton Report
32. Columbus Daily Sun March-April 1865
33. National Archives, Record Group 217, Microfiche # M1658, Fiche # 1, 2, & 3; Claim Number 15615
34. Dameron, 305
35. OR Series 1, Vol. 49, Part I, p. 929
36. Ibid., p. 1011-12
37. Ibid., p. 1041
38. Ibid., p. 1023
39. Columbus Daily Sun March 31, 1865
40. OR Series 1, Vol. 49, Part 2, p. 1193
41. Ibid., p. 1199
42. Columbus Daily Sun April 4, 1865
43. AA RH-46: SG 24912
44. Columbus Enquirer April 11, 1865; Russell County History, P. C-45; Columbus Enquirer February 5, 1965; Walker, p. 191 & 205
45. Russell County History, P. C-45
46. Ibid., p. C-45, F-154
47. Columbus Daily Sun April 8, 1865
48. OR Series 1, Vol. 31, part II, p. 660, 762; Vol. 52, Part II, p. 793
49. Columbus Daily Sun April 10, 1865
50. Columbus Daily Sun April 15, 1865
51. OR, Series 1, Vol. 49, Part 2, p. 1208
52. Ibid., p. 1212
53. Bragg, p. 107
54. Columbus Daily Sun, April 12, 1865
55. Ibid., April 15, 1865
56. OR Series 1, Vol. 52, Part 2 Chapter LXIV, "Supplements", p. 813
57. Turner, p. 233
58. Crumpton, p. 100
59. Greene Collection

60. Sisafikis; Dyer; Grant
61. Columbus Hospital Reports (April 16-22, 1865)
62. Hoole, p. 15
63. Nisbet, p. 256
64. Ibid. p. 257

CHAPTER 5

1. Turner, p. 233
2. Worsley, 294
3. Worsley, 294
4. Jones, Columbus State University Archives
5. Comer Diary, Columbus State University Archives
6. Stanton letter
7. Conzett, 76
8. Yeoman, p. 223
9. bid.
10. Yeoman, p. 223
11. Gilpin, p. 652
12. Barfield, p. 752
13. Gilpin, p. 652
14. Yeoman, p. 223
15. Worsley, p. 295
16. Yeoman, p. 224
17. Ibid., p. 225
18. Russell County History, p. C-46
19. Gilpin, p. 652
20. Michie, p. 167
21. ColumbusEnquirer, June 27, 1865; Martin, Part II, p. 179
22. Scott, p. 492
23. Ibid., p. 491
24. Wilson, Vol. 2, p. 259-60
25. Ibid., Vol. 2, 261
26. Gilpin, p. 653
27. Scott, p. 492
28. Hinricks Diary, April 16
29. OR Series 1, Vol. 49, Part 1, p. 493, Noble's Report
30. Harrington Diary, April 16, 1865
31. OR Series 1, Vol. 49, Part 1, p. 481- Winslow's Report
32. Hinricks Diary, April 16
33. OR Series 1, Vol. 49, Part 1, p. 363, Wilson's Report
34. Confederate Veteran April 1915, Vol. 23, p. 164
35. Wilson, Vol. 2, p. 261

36. Conzett, p. 77-78
37. Gilpin, p. 654
38. OR Series 1, Vol. 49, Part 1, p. 493, Noble's Report
39. Ibid.,
40. Hinricks Diary, April 16
41. Scott, p. 495
42. Jones, Columbus State University Archives
43. Scott, p. 496
44. Hinricks Diary, April 16
45. Ibid.
46. OR Series 1, Vol. 49, Part 1, p. 494, Noble's Report
47. Scott, p. 496
48. Ibid.
49. Harrington Diary, April 1865
50. Gilpin, p. 654
51. Barfield, p. 752
52. Ibid.
53. Scott, p. 497
54. Barfield, p. 752
55. Nisbet, p. 255
56. Gilpin, p. 654
57. Alphonza Jackson, Diary and Letters. Box 76-7, Microfilm Library, Georgia Department of Archives and History, quoted in Bragg, p. 107
58. Ibid.
59. OR Series 1, Vol. 49, Part 1, p. 495, Noble's Report
60. Siifakis, Bergeron, CS records
61. Confederate Veteran April 1915, Vol. 23, p. 164
62. Wilson, p. 263
63. Scott, p. 497
64. OR Series 1, Vol. 49, p. 399
65. Confederate Veteran April 1915, Vol. 23, p. 164
66. Greene Collection
67. Crumpton, p. 100
68. Ibid., p. 101
69. Scott, p. 499
70. Russell County History, p. C-46

71. Ibid.; Columbus Enquirer, June 27, 1865
72. OR Series 1, Vol. 49, Part 1, p. 499
73. Crumpton, p. 101
74. Harrington Diary, April 1865
75. Scott, p. 500
76. Ibid.; OR Series 1, Vol. 49, Part II, p.383
77. Harrington Diary, April 1865
78. Hinricks Diary, April 10, 1865
79. Scott, p. 500
80. Scott, p. 500; Turner, p. 233
81. Ibid.

CHAPTER 6

1. Gilpin, p. 654
2. Ibid.
3. OR Series 4, Vol. 3, pp. 718-20; Columbus Hospital Records (April 16-22, 1865)
4. Muscogee County Hospital records; Greene Collection
5. OR Series 1, Vol. 49, Part 1, p 407-409, Salter's Report
6. Columbus Sexton's Reports, Book E, 1853-1866, City Council records, City Clerk's Office
7. Greene Collection
8. Ibid.
9. Confederate Veteran April 1915, Vol. 23, p. 164
10. Barfield, p. 752
11. Navy Grey, p. 234
12. Greene Collection
13. Wilson, Vol. 2, p. 265
14. Hinricks Diary, April 17, 1865
15. Ibid.
16. Ibid.
17. Gilpin, p. 654
18. Columbus Enquirer, June 1, 1865; November 30, 1938; OR Series 1, Vol. 49, Part I, p. 478
19. Telfair, p. 137
20. Nisbet, p 255-56
21. Greene Collection
22. Charles Swift paper
23. OR Series 1, Vol. 49, Part II, p.383
24. OR Series 1, Vol. 49, Part 1, pp. 486-87, Winslow's report
25. Greene Collection: Coulter File
26. OR Series 1, Vol. 49, Part I, pp. 407-409 Salter's Report
27. Mitchell, p.192
28. Worsley, p. 297
29. Hinricks Diary & Harrington Diary, April 28; Greene Collection
30. National Archives, Record Group 217, Microfiche # M1658, Fiche # 1, 2, & 3; Claim Number 15615

31. Hinricks Diary, April 17
32. Harrington Diary, April 17
33. Wilson, p. 268
34. Hinricks Diary, April 18, 1865; OR Series 1, Vol. 49, Part 1, p. 407-09, Report of Surgeon F. Salter
35. Hinricks Diary, April 18, 1865
36. Wilson, Telfair, Greene Collection
37. Columbus Sexton's Reports, Book E, 1853-1866, City Council records, City Clerk's Office; OR, Series 1, Vol. 49, Part 2, P. 1271-72
38. Wilson, Vol. 2, p. 277
39. Ibid., p. 285-286
40. Telfair, p. 144
41. OR, Series 1, Part 2, p. 597

Appendix A

Accounts from the Official Records regarding Confederate flags captured by Union troops during the Battle of Columbus

1. **SGT L. Birdsall, B Company, 3rd Iowa Cavalry**, "captured the bearer and flag while my company was assailing the line of works on left of Summerville road. (Garrison flag)

2. **PVT Andrew W. Tibbets, I Company, 3rd Iowa Cavalry**, captured the bearer-a sergeant-and flag of Austin's battery, inside the line of works and to the right of the four-gun battery on the right of the enemy's line.

3. **PVT John Hays**, F Company, 4th Iowa Cavalry, "captured the standard and bearer, who tore it from the staff and tried to escape; he fired two shots from his revolver, wounding one man of my regiment at my side."

4. **CPL Richard I. Morgan,** A Company, 4th Iowa Cavalry, "I captured the standard and bearer in the first charge my company made, inside the line of works, the bearer contested with me for its possession."

5. **SGT Norman F. Bates,** E Company, 4th Iowa Cavalry, took a rebel and standard in the street three blocks from the bridge. (7th Alabama Cavalry)

6. **PVT Richard I. Cosgriff,** L Company, 4th Iowa Cavalry, on the west end of the bridge, "captured a standard and the bearer, having to knock him down with the butt of my gun before I could get possession of the flag."

7. **PVT John Kinney, L Company**, 4th Iowa Cavalry, captured a standard and bearer of Tenth Missouri Battery. "I had a tussle with the fellow to get the flag."

8. **PVT Edward J. Bebb,** D Company, 4th Iowa Cavalry, about 100 yards from the bridge and in the line of works, took a flag, the rebels near it running away before our men, leaving the flag.

Appendix B

Battle Analysis: The Battle of Columbus/ Girard
April 16-17, 1865

Overview:

Frequently overlooked as a battle of insignificant brutality and totally unnecessary, the Battle of Columbus, Georgia and Girard, Alabama was one of the last battles in the long and violent American Civil War (the last battle east of the Mississippi). The Union Cavalry Corps of Brevet Major General James H. Wilson attacked the composite remnants of both Alabama and Georgia troops commanded by Major General Howell Cobb.

The industrial center of Columbus, Georgia was a target in a series of planned attacks in a campaign that had begun that spring. Sweeping eastward across Alabama and Georgia to eliminate Confederate resistance, destroy materiel and industrial facilities, "Wilson's Raid" was a brilliant Union success; however, the victory was completely overshadowed by the surrender of General Lee's Army of Northern Virginia on April 9th and the crumbling of the Confederate government in Richmond.

On April 16, 1865, the Union cavalry forces commanded by James H. Wilson attacked the western earthwork defenses that guarded the Confederate industrial center of Columbus, Georgia. Wilson was attacking a region with severed lines of communications and he was uncertain of this rumored circumstance of Lee's surrender until days after the battle of Columbus.

"Wilson's Raid" into Georgia began weeks earlier as he swept the largest Cavalry force ever assembled southward from Tennessee in the spring of 1865. Wilson's massive raid first cracked the Confederate forces in Alabama with an attack on the cavalry forces of General Nathan B. Forrest at the Battle of Ebenezer Church on April 1, 1865.

Continuing his sweep southward, Wilson's Union cavalry then shattered resistance in Selma, Alabama on April 2nd. He then turned his forces eastward to the old Confederate capital of Montgomery, which surrendered without a fight on April 12th. As the demoralized Confederates fled into Georgia, hasty defenses were organized along the strategic bridges of the Chattahoochee River at Columbus, Georgia.

Columbus, Georgia was an invaluable Confederate commodity as the town was a large industrial center and second only to Richmond in its wartime industrial production. Columbus was a Confederate lifeline providing pistols, swords, bayonets, shoes, uniforms, tents, buckets, and a multitude of accoutrements throughout the war. It also served as a Naval port and shipbuilding facility.

Furthermore, Columbus served as the regional hub for cotton warehousing and transshipment via the Chattahoochee River, which empties southward into the Gulf of Mexico.

As survivors and refugees from Selma and Montgomery, Alabama fled eastward, they carried word of the impending arrival of Union forces. Confederate cavalrymen of General Abraham Buford's command clashed with Wilson's corps repeatedly in attempts to delay the massive thrust of the Union forces.

These brief delays were little more than a nuisance for Wilson's cavalry, but nonetheless, time was critical for the defenders of Columbus as Negroes were forced to dig a "tete de pont" at the northernmost wagon bridge across the Chattahoochee River. While two wagon bridges and one Railroad Bridge could support adequate conveyance of large forces, the upper (14th Street Bridge) connecting Girard (now Phenix City) Alabama with Columbus, Georgia was the focus of the Confederate defense.

For twenty miles above and below Columbus, the bridges at Columbus would have to be crossed or Wilson would be forced to build a pontoon bridge under fire of Confederate defenders who were determined to keep the Union raiders out of Georgia.

In the Official Records of the war maintained by the US government, the hostilities at Columbus have been expressed as a "battle," a "skirmish," and as an "action." Thus, the divisive debate concerning its actual prominence in history creates some controversy. With a full Cavalry Corps, Wilson's raiders pitted a possible force of nearly 13,000 men against an estimated Confederate force of 3,250 entrenched forces armed with 24 light pieces of light artillery in fortified positions.

The military objective of the Union force was to destroy the industrial capabilities present in Columbus, Georgia and continue to Macon. Meanwhile, the Confederate forces were determined to contain the Union marauders in Alabama (west of the Chattahoochee) and defend Columbus from their antagonists.

Wilson's Union Cavalry Corps was well equipped, well trained, readily supplied, and overall an extremely efficient organization. He and his unit commanders were also extremely efficient and experienced. One of Wilson's commanders was Emory Upton, who later wrote the US Army's basic manual of tactics and drill (Upton's Tactics), which served the Army for several decades following the Civil War.

Cobb's Confederate units were an ad-hoc organization that was neither well trained nor well organized. In fact, Cobb deferred overall command to the local defense force commander, Colonel Leon Von Zinken. While Von Zinken was an experienced and capable officer, he was forced to defend in place a large industrial complex housed within a city filled with inhabitants.

Furthermore, his units were remnants of their original organizations and the bulk of his men were reserve forces and either too young or too old for field service. Many men were members of the "Invalid Corps" and again ill-prepared to fight at all.

While Von Zinken did have elements of Buford's cavalry available, they were used to guard the flanks of the city to the north and south. Additionally, while the defenses contained fortified positions, the Confederates lacked a sufficient quantity of men to man them all.

Context: To place the combatant forces in proper context, the Union forces were far superior in both numbers and capabilities than their Confederate antagonists.

Yet, the Confederates did have forewarning of the impending arrival of their foe, and they were situated in a defensive position awaiting the imminent attack. For Wilson, the battle was yet another step in his overall campaign goal to sweep across to Macon, Georgia destroying all enemy resistance in the region.

Having attacked Selma, Alabama in a similar matter two weeks prior to this assault provided him with valuable insight as to how the Confederates would fight.

Again, after his victory at Selma, Wilson planned a similar assault on Montgomery, but the weakened forces and local civilians knew they could not withstand his forces, and they surrendered. Wilson may have hoped that Columbus would surrender as well, but he was just as prepared to fight if required.

In Columbus, the Confederates were determined to defend their resources and repel Wilson's assault, although they were all uncertain as to how they would do with such a beleaguered and ill-prepared composite organization.

Objectives: The combatants were as simple as attack (Union) and defend (Confederate). Both sides had plenty of time to prepare for battle and they both had adequate arms and ammunition. While the Confederates lacked enough men to fill all fortifications as desired, they consolidated their men into a "tete de pont" covering the northernmost wagon bridge and prepared them all for immediate destruction (upon command) by soaking them with turpentine and cotton.

This measure was to prevent the use of the bridges by the Union forces, and further north at West Point, Georgia, General Tyler was entrenched and prepared to defend the crossing at that location.

Wilson desired to capture the bridges at one locale or both, cross the Chattahoochee, and destroy the targets, eliminate all resistance, and continue across Georgia. Additionally, Von Zinken had organized his Confederate artillery into excellent positions, and apparently, the Confederates were confident in their ability to repel the attacking cavalry with long-range cannon-fire.

Additional factors: Several key personalities had a direct impact on both strategy and tactics. The Union force was an elite unit comprised of cavalrymen led by young generals that constantly pushed hard. Their aggressive nature coupled with innovation and initiative played key roles in their assault.

Likewise, but in the exact opposite context, the Confederate forces were led by aging leaders who displayed a lack of initiative and may have been suffering from a lack of self-confidence.

While both General Cobb and Buford were present in the defense of Columbus, they had suffered many defeats (professionally and personally) and their deferment of command to Von Zinken indicates a lack of commitment to the task at hand, and plausible deniability in case of failure.

However, in Buford's defense, his unit was nearly decimated when it reached Columbus and they were exhausted having skirmished with Wilson for days in attempts to delay him. Still, Cobb was the senior man present and his reputed weakness as a field commander was common knowledge throughout the Confederacy.

Alternatives: Viewing the objective of Columbus, Georgia as a target from the Union position in Montgomery, Alabama, Wilson had several options available. His forces would have to travel a great distance northward or southward and then eastward to West Point, Georgia (north) or Eufaula, Alabama (south).

Either course would take several more days of travel than a direct route, and there was a scarcity of habitation, food, forage, etc., in these rural regions to sustain his corps. Additionally, rapid and aggressive movement was the hallmark of cavalry operations, thus the direct route was the most desirable course of action.

Furthermore, the terrain (swampy lowlands and thick brush) channeled his movement along the main wagon road that led eastward through the small towns of Tuskegee and Crawford, Alabama.

When Wilson reached Columbus, he would have to perform a reconnaissance and make a detailed plan of attack based on the situation as presented. His Union force was well prepared for this action, and a siege of Columbus was not an option.

From the Confederate perspective, the hastily organized force had to prepare fortified defenses and capitalize on their strength using artillery. The Union force was much too large to engage beyond Columbus except for hit and run delaying actions performed by the Confederate cavalry.

The only real alternatives available involved specific placement of the limited manpower available to defend the city and its industries.

Confederate engineer, General Gilmer had constructed 10 forts to the east of Columbus in 1863, along the roads leading to Columbus, but most of these were too far (1-2 miles) beyond the river and the Confederates lacked adequate personnel to man them all. A series of defense lines placed in depth (exterior, intermediate and interior) could have been constructed (like Richmond), but the Confederates chose not to do this, citing a lack of personnel.

Area of Operations: The Confederates were established in defensive positions that consisted of several earthwork fortifications and connecting trench lines. All manned positions were within ½ mile west of Columbus, Georgia and oriented to defend the main wagon bridge (14th Street Bridge) in a "tete de pont."

These defensive positions were located on the Alabama side of the river and situated primarily to the northwest along the high (450 ft.) hills overlooking the river below. Several established dirt roads provided easy movement to the bridges that crossed the Chattahoochee River at Columbus.

In the north, the Summerville Road led downward to the main (14th Street Bridge) and in the south end of town, the Crawford Road led from Crawford, Alabama eastward to the Dillingham Street Bridge. These two bridges were about ¼ mile apart.

The Union cavalry approached Columbus from Crawford, Alabama, thus initially the lower (southern) Dillingham Street Bridge was in a direct path of assault. The lower bridge was lightly defended and prepped for destruction. Along the flanks (north and south) of the bridges, the Confederates placed cavalry to watch for attempted Union incursions outside of the defensive perimeter.

The Chattahoochee was too deep and too wide for a forced crossing; thus, Wilson would have to employ a pontoon bridge (he had adequate resources for this) or win a bridge to cross. There was another bridge (foot bridge) 3 miles north of Columbus at the Clapp's Factory. This bridge was prepared for immediate destruction also. The nearest wagon bridge beyond the three (2-wagon, 1 RR) at Columbus was 30 miles north at West Point, Georgia, and this location was also heavily defended.

Weather: Meteorological conditions on April 16-17 presented clear, comfortable springtime conditions for both sides (warm day, cool night). The most critical element that impacted both sides was the lunar conditions, which presented negligible luminosity. The level of darkness provided excellent conditions for a "fog of war," which either side was free to exploit.

Terrain: Columbus, Georgia lies in the heart of the Chattahoochee Valley where the Appalachian Mountain chain ends its southernmost point. This natural fall line geographically presents rolling hills to the north and west, and predominantly flat lands to the west and south. The maximum elevations in this area reach 450 feet.

As with most cities in 1865, much of the timber in the immediate area was removed for lumber and fuel, however, the woods along the riverbank, above and below the city were filled with pines and several varieties of small hardwood trees.

Size & composition:
- **Union:** 1 Cavalry Corps (13,000 men) & 1 4-gun battery (light howitzer), 250-wagon supply train w/ Pontoon Company
- **Confederate:** 3,250 (estimated) composite force- Alabama and Georgia reserves. 2 artillery batteries (+). Total of 24 light howitzers.

Technology:
- **Union:** Cannons, binoculars, Spencer rifles (repeaters).
- **Confederate:** Cannons, binoculars, several gunships in the river (Rams).

Logistics:

- **Union:** well-organized quartermaster train and self-sustaining forage capability.
- **Confederate:** Well-organized quartermaster depot, and warehouses filled with food, ammunition, and arms.

Command, Control, and Communications (C3):
- **Union:** Well-organized command & control system, bugles, employed scouts and informants (no telegraph).
- **Confederate:** Weak command & control system, bugles, drums (no telegraph).

Intelligence:
- **Union:** Employed reconnaissance scouts and informants as standard operational procedure (exploited Intel prior to commitment of forces).
- **Confederate:** Limited to local patrols (weak).

Doctrine & training:
- **Union:** Very strong, well honed, and experienced.
- **Confederate**: Weak, disorganized, and lacked confidence.

Condition & morale:
- **Union:** Strong, well-conditioned, and excellent morale.
- **Confederate:** Weak, numerous invalids, and poor morale.

Leadership:
- **Union:** Strong, experienced, and innovated.
- **Confederate:** Weak, lacked confidence in themselves, their men, and their mission.

Forces engaged in the Battle of Columbus/ Girard:
Union:
Cavalry Corps Commander- Brevet Major General

James H. Wilson

4th Division- Brevet Major General Emory Upton. 1st. Brigade. Brevet Brigadier General Edward F. Winslow. 3rd Iowa, Colonel John W. Noble. 4th Iowa, Lieutenant Colonel John H. Peters. 10th Missouri, Lieutenant Colonel Frederick W. Benteen. 2nd Brigade. Brevet Brigadier General Andrew J. Alexander. 5th Iowa, Colonel J. Morris Young. 1st Ohio, Colonel Beroth B. Eggleston. 7th Ohio, Colonel Israel Garrard. 4th United States Artillery, Battery I, Lieutenant George B. Rodney. **Confederate:** (estimated 2, 700 men and 24 cannons)

Commander- Major General Howell Cobb, Georgia State Reserves

(1st & 2nd Infantry Regiments- 1,600 men)

General Abraham Buford's 7th Alabama Cavalry (remnants-estimated 200 men)

Colonel Leon Von Zinken (overall commander)
Local Defense Force, Columbus, Georgia (estimated 600 men & 4 cannons)

Local Defense Force, Girard, Alabama (estimated 100 men)

Clanton's Alabama Artillery Battery (estimated 100 men and 10 guns)

Waddell's Alabama Artillery Battery (estimated 100 men and 10 guns)

Missions of opposing forces:

Union: Attack and destroy all Confederate resistance, war materiel, and war-making capabilities.
Confederate: Defend the city of Columbus; repel the Union attack, safeguard industrial facilities and materiel.
Initial disposition of forces:

Union: April 14-16, General Wilson spits his corps, sending the 1st Division under command of General LaGrange to assault West Point, Georgia in the north, while the remainder of his corps continues eastward to attack Columbus. Wilson's corps traveled along the Crawford Road and rotated units in the lead column. As it reached the outskirts of Girard, Alabama Confederate skirmishers were repelled and the Dillingham Street Bridge came into view. The column was halted and after a brief reconnaissance, a deliberate cavalry assault was directed against the Dillingham Street Bridge.

Confederate: All units were manning their defensive positions and cannons were prepared with grape and canister. As skirmishers placed to the east of Columbus fired the opening shots, the entire defenses braced to repel an imminent attack.

Opening moves: April 16, 2:00 pm. Recon by fire. Colonel Beroth B. Eggleston (lead unit in column) commanding the 1st Ohio regiment arrives on the outskirts of Girard, Alabama and battles Confederate skirmishers. He then occupies positions on the heights overlooking and east of the Dillingham Street Bridge, and he organizes for a charge.

General Emery Upton commanding the 4th Division watches from the heights as the 1st Ohio charges the bridge. Eggleston's men receive heavy artillery and small arms fire as they race towards the bridge.

As the 1st Ohio gets within a few hundred yards of the bridge, the Confederates fire the bridge, and Eggleston retreats to the heights where Upton has emplaced several of his cannons (Battery I, 4th US Artillery- Captain George B. Rodney) and exchanges fire with the Confederate artillery.

This duel provides Upton with information concerning the Confederate artillery positions, and his observations convince him to leave the bulk of his second brigade in that location, while he withdraws the rest of his division to circle around the Confederate positions to the west and north of

Girard.

5:00 pm. Redirect. General Upton sends a company commanded by Captain Young, of the Tenth Missouri Cavalry, to attempt the capture of the bridge at Clapp's factory, which upon their arrival, the Confederates torched the bridge and prevented a crossing at that location.

Meanwhile, Upton performs a personal reconnaissance and gains information from a local "citizen" about the fortified positions located on Ingersoll Hill along the Summerville Road (northwest heights above the 14th Street bridge).

Upton organizes his forces to assault these positions. General Winslow's brigade is tasked to provide pickets and to march southward perpendicular to the Summerville Road.

Major Phases:

08:00 pm. Assault. Two companies of the Fifth Iowa Cavalry, under Captain Lewis, spearheads the assault with his pickets. General Winslow moves his brigade forward with dismounted companies in the front lines and several mounted companies held behind them as a reserve force with the hope of punching through the defense works and charging straight down to the bridge.

09:00 pm. Breach. As darkness enveloped the scene, the assault was placed into full motion with the addition of six more companies (Third Iowa Cavalry, commanded by Colonel Noble). The Confederate front line was soon breached, and General Upton sent two companies of the Tenth Missouri Cavalry to charge down to the bridge. These companies passed right through the enemy's lines and charged straight down the Summerville road.

Apparently, the Confederates mistook the Union cavalry as friendly troops and allowed them to pass through the lines unmolested right down to the bridge. Captain McGlasson (leading the Missouri cavalry) secured the bridge on the west side of the bridge (Girard). While these troops had achieved success, they were surrounded by the enemy.

Thus, McGlasson ordered his column to return up the hill, to friendly lines. On the heights above, the Confederate artillery continued to fire at the advancing lines of Union troops with grapeshot.

Due to the darkness and confusion, the artillery was aimed too high and the Union lines easily continued their advance. The Third Iowa soon took the forts that guarded the heights. The Fourth Iowa Cavalry (dismounted and led by Captain Lot Abraham), moving downhill and to the right of the Third Cavalry also breached the defense works in his front and captured the artillery therein. Continuing the fight, Abraham ordered his men to charge the bridge, which was now filled with retreating Confederates. Thus, both Union and Confederates raced into Columbus and the artillery on that side of the river could not fire, lest they kill their comrades, nor could they fire the bridge.

10:00 pm. Consolidation. Once Abraham's men reached Columbus, a Confederate rally was attempted, but rapidly quelled by the Union cavalrymen. In this last vain attempt to halt the Union assault, Colonel Charles Lamar and several others of the Confederate command were killed in action. This represented the last remnant of any Confederate organized effort except for a hasty withdrawal led by General Cobb. Ordering the bulk of the Georgia State troops onto a train, Cobb ordered a retreat.

Thus, Cobb and at least 600 men of the Georgia line retreated easterly via the rails towards Macon, Georgia. The entire scene turned into a mass of confusion with Confederates fleeing Columbus and surrendering to the victorious Union cavalrymen. Within the fortified Confederate defense works, the Union troops rounded up prisoners, while below in Columbus, the Union spearhead rushed through the town capturing their enemy as they fled the scene.

Outcome: April 17, 1865. The new day brought the remaining elements of General Wilson's Corps, who set his men to work destroying the Confederate industrial capabilities. The Columbus Naval Yard, the CSS Jackson, and all the various arms and ammunition works were all destroyed and burned. General Wilson and his Union cavalry had achieved a total success.

All the Confederate forces that did not escape were taken into captivity (1,500) and Columbus was in Union hands. Actual casualty figures vary, but the Union suffered approximately 7 KIA and 56 WIA (most died of post-wound complications), while the Confederates suffered at hundreds of casualties. Wilson also destroyed 125,000 bales of cotton, which was very valuable to the Confederacy and he continued his march towards Macon.

In the afternoon of April 16, Wilson's cavalry also achieved their objective at West Point, thus the Union campaign was a total success. General Cobb surrendered his remaining forces to General Wilson on April 20th at Macon, Georgia.

Summary of events: Wilson achieved success primarily due to his exploitation of the "fog of war," and the tenacity of his forces. While the Confederates made numerous mistakes that provided the Union forces an easy victory, Wilson's men deserve credit for capitalizing on them.

The Confederates lacked a plan of recourse in case their lines were breached, their leadership was weak, command & control was weak, the artillery fired too high. While the Confederate defense was filled with weakness, the lack of strong Confederate leadership, the strength of Union leadership, and the opportunities afforded in the "fog of war" provided an easy victory for the Union.

Relevance of this battle to current operations: The relevance of this battle to current operations is primarily the exploitation of darkness. Night operations provide great opportunities to he who "owns the night." With the current technologies afforded the modern warrior with night vision devices and Infrared lenses, the "fog of war" is virtually lifted and open for total exploitation.

Additionally, the force multipliers of good training and strong leadership are evident in this battle, and provide a relevant example of their importance on the battlefield. The Union forces under Wilson were well trained, aggressive, and not afraid to use innovation and decisive actions to achieve their mission. While current technologies employed by the Army of today could help prevent the confusion that was prevalent in this battle, we still require leaders who can make decisions quickly, and exploit opportunities decisively.

Who won? The Union forces led by General James Wilson were the obvious victors. Generals Wilson and Upton are clearly winners at the higher echelons (strategy and tactics), while both Captains McGlasson and Abraham provided heroic leadership and were clearly winners in the lower echelons.

Who lost?
The Confederate forces led by General Howell Cobb were the obvious losers. General Cobb and Colonel Von Zinken are clearly losers at the higher echelons (strategy and tactics), while Clanton and Waddell's artillery work is questionable.

What were the constants that affected the outcome?
Training is the key. For the Union, good training supported their mission very well and served as a force multiplier, while the

Confederates experienced the exact opposite due to their lack of training/ coherent planning and actions.

Morale may have been a constant as well since the Confederates were greatly impacted knowing that the South was losing the war. Again, for the Union, their morale was high and this constant probably supported their efforts as well.

Principles of War:

Objective: Clearly defined on both sides, but the Union forces made certain that all levels understood both the goals and the objectives. The commander's intent was to capture the bridge and they accomplished their mission.

Offensive: Completely exploited and strategically employed by the Union forces.

Mass: The Confederate forces were massed well, but they lacked a coherent plan. The Union forces employed massed forces effectively while maintaining adequate reserves for flexible maneuver.

Economy of force: Again, the Union capitalized on this principle, yet Cobb managed to save some reserves by retreating.

Maneuver: Expertly employed by all Union forces, and not used at all by the Confederates (partial cause of failure).

Unity of command: Excellent within the Union forces, and poorly arranged by the Confederates (partial cause of failure).

Security: Lack of adequate security within the Confederate ranks (partial cause of failure).

Surprise: Absolutely exploited by the Union forces (force multiplier and greatly contributed to success).

Simplicity: The Confederate defense was too simple and the Union plan proved adequate to the task at hand.

Tenets of Military Operations:

Initiative: This was a great force multiplier for the

Union forces and expertly exercised. The lack of initiative within the Confederate forces contributed to their failure.

Agility: The inherent agility of the Union cavalry contributed immensely to their flexible maneuver and victory.

Depth: Had the Confederates been able to build their defense in-depth the Union may have failed to breach the line.

Synchronization: Excellent application employed by the Union forces, probably achieved by the mobility and rapidity of communications between mounted units. Totally absent within the Confederate lines and contributed to their failure.

Versatility: A great force multiplier for the Union forces and readily achieved by their ability to mount and dismount.

Battlefield Operating Systems (BOS):

Intelligence: Readily available to the Union forces through reconnaissance, informants, and actions conducted during daylight hours. The Confederates were limited in this regard due to their isolation within their entrenchments.

Maneuver: The greatest strength of cavalry force is their mobility combined with maneuver and Wilson's Union forces employed it skillfully. Again, the Confederates isolated themselves within the entrenchments and failed to plan for adequate reactions.

Fire support: Both sides had cannons for fire support, and this should have been the greatest strength of the Confederate forces, unfortunately they failed to employ it correctly.

Air defense: Not applicable in this battle.

Mobility: The maneuver of the cavalry forces combined within this BOS was key to Union success.

Survivability: Both sides were logistically prepared for this BOS, but the Confederates weak leadership and planning made survivability as nearly non-existent as evidenced in their rapid losses on the battlefield.

Logistics: Both sides enjoyed excellent logistical support, and as planned, Wilson's victory helped destroy the Confederates logistical capabilities and resources.

Battle command: Exercised with great efficiency within the Union chain of command and contributed to their success. In just the opposite manner for the Confederates, the lack of battle command directly contributed to their loss.

Appendix C

Image Credits

Library of Congress- LOC

Library of Congress-Map Division- LOC MD

George Greene Collection- GC

Columbus State University- CSU

Author's Collection- AC

Page #	Title	Source
14	MG James H. Wilson	LOC
17	Union Cavalry Officer	LOC
19	Spencer Rifle graphics	AC
20	Union Cavalry Troopers	LOC
21	Wilson's camp in 1865	LOC MD
23	LTG Richard Taylor	GC
25	LTG Nathan B. Forrest	GC
26	BG James Chalmers	GC
27	BG William Jackson	GC
29	BG Abraham Buford	GC

Page #	Title	Source
30	LTG Forrest's Staff	GC
31	BG Long & BG McCook	GC
33	BG James Clanton	GC
34	BG Emory Upton	LOC
35	Philip D. Roddey	GC
36	1st Sergeant John Gammill	GC
39	BG Andrew Alexander	GC
40	Ebenezer Church Map	LOC MD
43	Selma Battle Map	LOC MD
50	Wilson's Raid Map	LOC MD
51	Colonel Robert Minty	GC
59	Columbus Map	GC
79	CSS Jackson	GC
82	MG Howell Cobb	GC
87	CPT Theodore Moreno	GC

88	Fortifications	GC
92	CPT John Pemberton	GC
108	Forts at Columbus Map	LOC MD
113	COL Von Zinken letter	GC
118	Captain Richard Bellamy and Major James Waddell	GC
127	Union Flags	AC
128	Confederate Flags	AC
134	Defensive Network Map	LOC MD
139	COL Morris Young	GC
143	COL Frederick Benteen	GC
147	COL Edward Winslow	GC
149	CPT Nathan Clanton	GC
151	Cavalrymen by Alfred Waud	LOC
161	Captain Lot Abraham	GC
162	PVT Andrew Tibbets	GC
170	SGT Norman Bates	GC
173	Battle Map of Columbus	AC
198	MG Wilson and Staff	LOC

Appendix D

Select Unit Histories

Within the *Official Records of the War of the Rebellion* (1880 - 1901), Marcus J. Wright's *List of Field Officers, Regiments, and Battalions in the Confederate States Army, 1861-1865* (1912), Civil War Soldier Records available in the National Archives, and the National Park Service, *Soldiers and Sailors database*, one can find the units that participated in key battles. The unit histories below are derived from these files.

Additional historical data available in local City and County Archives also provide key details (such as the CSA Hospital Records housed in Columbus, Georgia) and can be invaluable in piecing together details regarding battles.

The following select unit histories, Union and Confederate, of the Cavalry combatants at the Battle of Columbus provide an overview of key dates, events, and a brief service history of each unit. These units were selected based upon the fact that members of these units fought at the Battle of Columbus, and unit members appear on the casualty rolls.

3rd Regiment, Iowa Cavalry

Overview:
    Organized at Keokuk August 30 to September 14, 1861. Moved to Benton Barracks, Mo., November 4-6, and duty there till February 4, 1862. (Cos. "E," "F" "G" and "H" detached to Jefferson City, Mo., December 12,

1861, and duty in Northern and Southern Missouri till July, 1863. See service following that of Regiment.) Cos. "A," "B," "C," "D," "I," "K," "L" and "M" moved to Rolla, Mo., February 4-6, 1862. (Cos. "I" and "K" detached to garrison, Salem, Mo., February 11, 1862. Scout to Mawameck February 12. Expedition to Mt. Vernon February 18-19. Action at West Plains February 20. Scouting after Coleman's guerillas till April. Actions near Salem February 28 and March 18. Rejoin Regiment near Forsythe April, 1862.) Regiment march to join General Curtis February 14-18. (Co. "L" detached at Springfield, Mo.) Attached to Curtis' Army of Southwest Missouri, Dept. of Missouri, February to May, 1862. 3rd Brigade, 1st Division, Army of Southwest Missouri, to July, 1862. District of Eastern Arkansas, Dept. of Missouri, to October, 1862. 3rd Brigade, 4th Division, District of Eastern Arkansas, to December, 1862. 2nd Brigade, Cavalry Division, District of Eastern Arkansas, Dept. of Tennessee, to January, 1863. 2nd Brigade, 2nd Cavalry Division, 13th Corps, Dept. of Tennessee, to April, 1863. 2nd Brigade, Cavalry Division, District of Eastern Arkansas, Dept. of Tennessee, to June, 1863. Bussy's Cavalry Brigade, Herron's Division, Dept. of Tennessee, to August, 1863. Reserve Cavalry Brigade, Army of Arkansas, to January, 1864. 1st Brigade, 1st Division, 7th Army Corps, Dept. of Arkansas, to May, 1864. 2nd Brigade, Cavalry Division, 16th Corps, Dept. of Tennessee, to June, 1864. 2nd Brigade, 2nd Cavalry Division, District of West Tennessee, to December, 1864. 2nd Brigade, Cavalry Division, District of West Tennessee, to February, 1865. 1st Brigade, 4th Division, Wilson's Cavalry Corps, Military Division Mississippi, to June, 1865. District of Georgia to August, 1865.

Service:

Expedition to Fayetteville, Ark., February 22, 1862. Battles of Pea Ridge March 6-8. (Cos. "D" and "M" escort prisoners to Rolla, Mo., March 12-31.) March to Batesville via Cassville, Forsythe, Osage and West Plains April 6-May 1. (Cos. "L" and "M" detached at Lebanon, Mo., operating against guerillas till November, 1862; then join Cos. "E," "F," "G" and "H"). (Co. "D" guard train to Rolla, Mo., May 25 to June 20.) Action at Kickapoo Bottom, near Sylamore, May 29. Sylamore May 30. Foraging and scouting at Sulphur Rock June 1-22. Waddell's Farm, Village Creek, June 12. March from Batesville to Clarendon on White River June 25-July 9. Waddell's Farm June 27 (Co. "K"). Stewart's Plantation, Village Creek, June 27. Bayou Cache July 6 (Co. "I"). Hill's Plantation, Cache River, July 7. March to Helena July 11-14. Duty there and scouting from White River to the St. Francis till June, 1863. Expedition from Clarendon to Lawrenceville and St. Charles September 11-13, 1862. LaGrange September 11. Marianna and LaGrange November 8. Expedition to Arkansas Post November 16-21. Expedition to Grenada, Miss., November 27-December 5. Oakland, Miss., December 3. Expedition up St. Francis and Little Rivers March 5-12, 1863 (Detachment). Expedition to Big and Little Creeks and skirmishes March 6-10. Madison, Ark., March 9 (Detachment). Madison, Ark., April 14 (Detachment). LaGrange May 1. Polk's Plantation, Helena, May 25. Moved to Vicksburg, Miss., June 4-8. Siege of Vicksburg June 8-July 4. Advance on Jackson, Miss., July 5-10. Near Clinton July 8. Siege of Jackson July 10-17. Near Canton July 12. Canton, Bolton's Depot and Grant's Ferry, Pearl River, July 16. Bear Creek, near Canton, July 17. Canton July 18. At Flowers' Plantation till August 10. Raid from Big Black on Mississippi Central Railroad and to Memphis, Tenn., August 10-22. Payne's Plantation, near Grenada, August 18. Panola August 20. Coldwater August 21. Moved to

Helena, Ark., August 26; thence moved to Little Rock, arriving October 1. Duty at Berton, Ark., October 1 to December 20. Expedition to Mt. Ida November 10-18. Near Benton December 1. Expedition to Princeton December 8-10. Ordered to Little Rock December 20. Regiment Veteranize January 5, 1864. Veterans on furlough January 6 to February 5. At St. Louis, Mo., February 6 to April 26. Ordered to Memphis, Tenn., April 26. Operations against Forest May to August. Sturgis' Expedition to Guntown, Miss., June 1-13. Near Guntown June 10. Ripley June 11. Smith's Expedition to Tupelo, Miss., July 5-21. Saulsbury July 2. Near Kelly's Mills July 8. Cherry Creek July 10. Huston Road July 12. Okolona July 12-13. Harrisburg, near Tupelo, July 14-15. Old Town or Tishamingo Creek July 15. Ellistown July 16 and 21. Smith's Expedition to Oxford, Miss., August 1-30. Tallahatchie River August 7-9. Holly Springs August 8. Hurricane Creek and Oxford August 9. Hurricane Creek August 13, 14 and 19. College Hill August 21. Hurricane Creek August 22. Repulse of Forrest's attack on Memphis August 21 (Detachment). Moved to Brownsville, Ark., September 2. Campaign against Price in Arkansas and Missouri September-November. Independence, Big Blue and State Line October 22. Westport October 23. Battles of Charlot, Marias des Cygnes, Mine Creek, Little Osage River October 25. White's Station, Tenn., December 4 (Detachment). Grierson's Raid from Memphis on Mobile & Ohio Railroad December 27, 1864, to January 6, 1865 (Detachment). Near White's Station December 25. Okolona December 27. Egypt Station, Miss., December 28. Mechanicsburg January 3, 1865. At the Pond January 4. Moved from Vicksburg, Miss., to Memphis, Tenn.; thence to Louisville, Ky., January 6-15, 1865, and rejoin Regiment. Regiment at St. Louis, Mo., and Louisville, Ky., till February, 1865. Moved to Chickasaw, Ala.; Wilson's Raid to Macon, Ga., March

22-April 24. Montevallo March 31. Six-Mile Creek March 31. Maplesville April 1 (Co. "L"). Ebeneezer Church, near Maplesville, April 1. Selma April 2. Fike's Ferry, Cahawba River, April 7 (Co. "B"). Montgomery April 12. Columbus, Ga., April 16. Capture of Macon April 20. Duty at Macon and at Atlanta, Ga., till August. Mustered out August 9, 1865.

Regiment lost during service 5 Officers and 79 Enlisted men killed and mortally wounded and 4 Officers and 230 Enlisted men by disease. Total 318.

Companies "E," "F," "G" and "H" ordered to Jefferson City, Mo., December 12, 1861. Attached to Army of Southwest Missouri to February, 1862. District of North Missouri to August, 1862. District of Southwest Missouri to November, 1862. Cavalry Brigade, District of Southeast Missouri, to June, 1863. Reserve Cavalry Brigade, Army of Southeast Missouri, to August, 1863. Reserve Brigade, 1st Cavalry Division, Arkansas Expedition, to October, 1863.

Service:
Engaged in operations against guerillas about Booneville, Glasgow, Fulton and in North Missouri at Lebanon, and in Southwest Missouri covering frontier from Iron Mountain to Boston Mountains till June, 1863. Companies "L" and "M" joined November, 1862. Actions at Florida, Mo., May 22, 1862. Salt River, near Florida, May 31. Boles' Farm, Florida, July 22 and 24. Santa Fe July 24-25. Brown Springs July 27. Moore's Mills, near Fulton, July 28. Kirksville August 26. Occupation of Newtonia December 4. Hartsville, Wood's Fork, January 11, 1863. Operations against Marmaduke April 17-May 2. Cape Girardeau April 26. Near Whitewater Bridge April 27. Castor River, near Bloomfield, April 29. Bloomfield April 30. Chalk

Bluffs, St. Francis River, April 30-May 1. Davidson's march to Clarendon, Ark., August 1-8. Steele's Expedition to Little Rock August 8-September 10. Reed's Bridge or Bayou Metoe August 27. Shallow Ford, Bayou Metoe, August 30. Bayou Fourche and capture of Little Rock September 10. Rejoined Regiment at Little Rock October 1, 1863.

4th Regiment, Iowa Cavalry

Overview:

Organized at Camp Harlan, Mount Pleasant, September to November, 1861. Companies muster in "A," "E" and "F" November 23, "B," "C," "D," "I," "K" and "M" November 25, "G" November 27, "L" December 24, and "H" January 1, 1862. Duty at Camp Harlan till February, 1862. 1st Battalion moved to St. Louis, Mo., February 26, 2nd Battalion February 28 and 3rd Battalion March 3, 1862. At Benton Barracks, Mo., till March 10. Ordered to Rolla, Mo., March 10; thence to Springfield, Mo., and duty there till April 14. Attached to 2nd Division, Army of Southwest Missouri, Dept. of Missouri, to July, 1862. District of Eastern Arkansas, Dept. of Missouri, to December, 1862. 2nd Brigade, 1st Cavalry Division, District of Eastern Arkansas, Dept. of Tennessee, to January, 1863. 2nd Brigade, 2nd Cavalry Division, 13th Corps, Dept. of Tennessee, to May, 1863. Unattached, 15th Army Corps, Army of Tennessee, to August, 1863. Winslow's Cavalry Brigade, 17th Corps, to May, 1864. 2nd Brigade, 1st Cavalry Division, 16th Corps, to July, 1864. 2nd Brigade, 2nd Cavalry Division, District of West Tennessee, to November, 1864. 1st Brigade, 4th Division, Wilson's Cavalry Corps, Military Division Mississippi, to December, 1864. 2nd Brigade, Cavalry Division, District of West

Tennessee, to February, 1865. 1st Brigade, 4th Division, Cavalry Corps, Military Division Mississippi, to June, 1865. Dept. of Georgia to August, 1865.

Service:

Expedition to Salem, Mo., March 12-19, 1862 (Cos, "F" and "L"). Ordered to join Curtis at Batesville, Ark., April 14. Skirmish at Nitre Cave, White River, April 18 (Detachment Cos. "G" and "K"). Talbot's Farm, White River, April 19 (Detachment Cos. "E," "F," "G" and "K"). Skirmish, White River, May 6. Little Red River June 5. (Co. "F" detached for duty with Chief Commissary and as provost guard at Helena, Ark., May, 1862, to April, 1863.) Mt. Olive June 7, 1862 (Co. "F"). Gist's Plantation July 14, 1862 (Co. "F"). March to Helena, Ark., June 11-July 14. Duty at Helena till April, 1863. Polk's Plantation September 20, 1862 (Detachment Co. "D"). Expedition from Helena to LaGrange September 26 (2 Cos.). Jones' Lane or Lick Creek October 11 (Detachment Cos. "A," "G" and "H"). Marianna and LaGrange November 8. Expedition from Helena to Arkansas Post November 16-21, and to Grenada, Miss., November 27-December 5. Oakland, Miss., December 3. Expedition to Big and Little Creeks March 6-12, 1863. Big Creek March 8. St. Charles and St. Francis Counties April 8. Moved to Milliken's Bend, La., April 28-30. Reconnaissance to Bayou Macon May 1-4. March to New Carthage May 5-8. (Co. "G" detached on courier duty at Young's Point, La., during May.) Fourteen-Mile Creek May 12-13. Mississippi Springs May 13. Hall's Ferry May 13 (Detachment). Baldwyn's Ferry May 13 (Detachment). Jackson May 14. Haines Bluff May 18 (Co. "B"). Siege of Vicksburg, Miss., May 18-July 4. Engaged in outpost duty against Johnston between Big Black and Yazoo Rivers. Mechanicsburg May 24 and 29. Expedition from Haines Bluff to Satartia and Mechanicsville June 2-8

(Detachment) Barronsville June 18. Bear Creek or Jones' Plantation June 22 (Cos. "A," "F," "I" and "K"). Big Black River, near Birdsong Ferry, June 22 (Detachment). Hill's Plantation, near Bear Creek, June 22. Messenger's Ferry, Big Black River, June 26. Advance on Jackson July 5-10. Siege of Jackson July 10-17. Near Canton July 12. Bolton's Depot July 16. Bear Creek, Canton, July 17. Canton July 18. Raid from Big Black on Mississippi Central Railroad and to Memphis, Tenn., August 10-22. Payne's Plantation, near Grenada, August 18. Panola August 20. Coldwater August 21. Expedition to Yazoo City September 21-October 1 (Detachment). Brownsville September 28. Morris Ford, near Burton, September 29. Expedition toward Canton October 14-20. Brownsville October 15. Canton Road, near Brownsville, October 15-16. Near Clinton and Vernon Cross Roads October 16. Bogue Chitto Creek October 17. Robinson's Mills, near Livingston, October 17. Louisville Road, near Clinton and Brownsville, October 18. Expedition to Natchez December 4-17 (Detachment Cos. "C," "H," "I," "K," "L" and "M"). Near Natchez December 7. Meridian Campaign February 3-28, 1864. Big Black River Bridge February 3. Raymond Road, Edwards Ferry, Champion's Hill, Baker's Creek and near Bolton's Depot February 4. Jackson and Clinton February 5. Brandon February 7. Morton February 8. Meridian February 9-13. Hillsborough February 10. Tallahatta February 13. Meridian February 14. Near Meridian February 19. Veterans on furlough March 4 to April 24. Reported at Memphis, Tenn., April 24. Non-Veterans at Vicksburg, Miss., till April 29; then moved to Memphis. Sturgis' Campaign against Forrest April 30-May 12. Sturgis' Expedition to Guntown, Miss., June 1-13. Ripley June 7. Brice's Cross Roads, near Guntown, June 10. Ripley June 11. Smith's Expedition to Tupelo, Miss., July 5-21. Near Ripley July 7. Cherry Creek July 10. Plenitude

July 10. Harrisburg Road July 13. Tupelo July 14-15. Old Town or Tishamingo Creek July 15. Smith's Expedition to Oxford, Miss., August 1-30. Tallahatchie River August 7-9. Hurricane Creek and Oxford August 9. Hurricane Creek August 13, 14 and 19. College Hill August 21. Oxford August 22. (Forrest's attack on Memphis August 21-Co. "G.") Moved to Little Rock, Ark., September 2-9. Campaign against Price in Arkansas and Missouri September 17-November 30. Moved to Batesville and Pocahontas, Ark.; thence to Cape Girardeau, St. Louis, Jefferson City and Independence, Mo., Trading Post and Fort Scott, Kansas, Pea Ridge and Fayetteville, Ark., Tahlequah and Webber's Falls, Ind. Ter., returning via Pea Ridge, Springfield and Rolla to St. Louis. Engaged at Brownsville September 28. Morris Bluff September 29 (Co. "D"). Little Blue October 21. Independence October 22. Westport, Big Blue and State Line October 23. Trading Post October 25. Marias des Cygnes, Osage, Mine Creek October 25. Charlotte Prairie October 25. At St. Louis till December 9; then at Louisville, Ky., till February, 1865. (A detachment at Memphis, Tenn., September 1 to December 20, 1864. Scout near Memphis November 10. Skirmish on Germantown Pike, near Memphis, December 14, Detachments of Cos. "A" and "B." Grierson's Raid on Mobile & Ohio Railroad December 21, 1864, to January 5, 1865. Okolona, Miss., December 27, 1864. Egypt Station December 28. Franklin January 2, 1865. Rejoined Regiment at Louisville, Ky., January 15, 1865.) Dismounted men of Regiment moved from Memphis, Tenn., to Louisville, Ky., January 2, 1865. Moved to Gravelly Springs, Ala., February, 1865, and duty there till March 20. Expedition to Florence March 1-6. Wilson's Raid to Macon, Ga., March 20 to May 10. (Co. "G" escort to General Upton, Commanding Division.) Montevallo March 30. Near Montevallo March 31. Six-Mile Creek March 31.

Ebenezer Church April 1. Selma April 2. Fike's Ferry, Cahawba River, April 7. Wetumpka April 13. Columbus, Ga., April 16. Capture of Macon April 20. Duty at Macon and Atlanta, Ga., till August. Mustered out at Atlanta August 10, 1865, and discharged at Davenport, Ia., August 24, 1865.

Regiment lost during service 4 Officers and 51 Enlisted men killed and mortally wounded and 5 Officers and 194 Enlisted men by disease. Total 254.

5th Regiment, Iowa Cavalry

Overview:

Organized as Curtis Horse by order of General Fremont. Cos. "A," "B," "C" and "D" organized at Omaha, Neb., September 14 to November 13, 1861; "E" at Dubuque, Ia.; "F" in Missouri, as Fremont Hussars, October 25, 1861; "H" at Benton Barracks, Mo., December 28, 1861; "G," "I" and "K" as 1st, 2nd and 3rd Independent Companies, Minnesota Cavalry, at Fort Snelling, Minn., October 29 to December 20, 1861; "L" as Naughton's Irish Dragoons at Jefferson City, Mo., and "M" as Osage Rifles at St. Louis, Mo., November 1, 1861. Duty at Benton Barracks, Mo., till February, 1862. Moved to Fort Henry, Tenn., February 3-11. Patrol duty during battle of Fort Donelson. Expedition to destroy railroad bridge over Tennessee River February 14-16. Attached to Dept. of the Tennessee, Unassigned, to November, 1862. District of Columbus, Ky., 13th Corps, Dept. of Tennessee, to December, 1862. District of Columbus, 16th Corps, to June, 1863. 1st Brigade, 2nd Division, Cavalry Corps, Army of the Cumberland, to August, 1863. 3rd Brigade, 2nd Division, Cavalry Corps, Cumberland, to November, 1863. 1st Brigade, 2nd

Division, Cavalry Corps, Cumberland, to April, 1864. 1st Brigade, 3rd Division Cavalry Corps, Cumberland, to November, 1864. 2nd Brigade, 6th Division, Wilson's Cavalry Corps, Military Division Mississippi, to December, 1864. 1st Brigade, 6th Division, Cavalry Corps, Military Division Mississippi, December, 1864. 2nd Brigade, 6th Division, Cavalry Corps, Military Division Mississippi, to February, 1865. 2nd Brigade, 4th Division, Wilson's Cavalry Corps, to June, 1865. Dept. of Georgia to August, 1865.

Service:
Garrison duty at Forts Henry and Heiman till February 5, 1863. Skirmish, Agnew's Ferry, March 25, 1862 (Detachment of Co. "K"). Moved to Savannah, Tenn., March 28-April 1 (Cos. "G," "I" and "K"). Moved toward Nashville, Tenn., repairing roads and erecting telegraph lines April 3-6. Advance on and siege of Corinth, Miss., April 29-May 30. Acting as escort to Telegraph Corps. Expedition from Trenton to Paris and Dresden May 2-9. Dresden May 5. Lockridge's Mills May 5. Occupation of Corinth May 30. Pursuit to Booneville May 31-June 12. Designated 5th Iowa Cavalry June, 1862. Duty at Humboldt, Tenn., till August. Companies "G," "I" and "K" rejoin Regiment. Paris, Tenn., March 11, 1862 (1st Battalion). Expedition to Paris March 31-April 2 (Co. "F"). Near Fort Donelson August 23 (Detachment). Fort Donelson August 23. Cumberland Iron Works August 26. Expedition to Clarksville September 5-10. New Providence September 6 (Cos. "G," "I" and "K"). Clarksville September 7. Operations about Forts Donelson and Henry September 18-23. Near Lexington Landing October 1 (Detachment). Scout toward Eddyville October 29-November 10 (Cos. "G," "I" and "K"). Garrettsburg, Ky., November 6. Expedition from Fort Heiman December 18-28 (Cos. "G," "I" and "K"). Waverly

January 16, 1863. Cumberland Iron Works, Fort Donelson, February 3, 1863. Moved to Fort Donelson February 5, and duty there till June 5. Destruction of Bridge, Mobile & Ohio Railroad, February 15. Paris, Tenn., March 14. Waverly April 10 (Detachment). Stewartsborough April 12 (1 Co.). Moved to Murfreesboro and Nashville, Tenn., June 5-11. Scout on Middletown and Eaglesville Pike June 10. Expedition to Lebanon June 15-17. Lebanon June 16. Middle Tennessee (or Tullahoma) Campaign June 23-July 7. Guy's Gap, Fosterville, June 25. Fosterville June 27. Expedition to Huntsville July 13-22. Moved to McMinnsville September 6-8, and operating against Guerillas till October. Wartrace September 6. Operations against Wheeler and Roddy September 30-October 17. Garrison Creek near Fosterville October 6. Wartrace October 6. Sugar Creek October 9. Tennessee River October 10. At Maysville till January, 1864. Expedition from Maysville to Whitesburg and Decatur November 14-17, 1863, to destroy boats on the Tennessee River. Outpost duty on line of the Tennessee River, from south of Huntsville to Bellefonte November and December. Veteranize January, 1864. On Veteran furlough January 7 to April 24. Non-Veterans at Nashville, Tenn., till May. Companies "G," "I" and "K" detached February 25, 1864, and designated Brackett's Battalion, Minnesota Cavalry. Moved from Nashville to Pulaski and guard Nashville & Decatur Railroad till July. Moved to Decatur July 5. Rousseau's Raid from Decatur on West Point & Montgomery Railroad July 10-22. Near Coosa River July 13. Ten Island Ford, Coosa River, July 14. Chehaw Station, West Point & Montgomery Railroad July 18. Siege of Atlanta July 22-August 25. McCook's Raid on Atlanta & West Point Railroad July 27-31. Lovejoy Station July 29. Clear Creek July 30. Near Newnan August 15. Flank movement on Jonesboro August 25-30. Flint River

Station August 30. Jonesboro Aug. 31-September 1. (5th Iowa Infantry consolidated with Regiment as Companies "G" and "I" September 1, 1864.) Lovejoy Station September 2-6. Operations against Hood in North Georgia and North Alabama September 29-November 3. Camp Creek September 30. Sweetwater and Noyes Creek, near Powder Springs October 1-3. Van Wert October 9. Nashville Campaign November-December. Columbia, Duck River, November 24-27. Crossing of Duck River November 28. Columbia Ford November 28-29. Battle of Nashville December 15-16. Pursuit of Hood December 17-28. Franklin and West Harpeth River December 17. Spring Hill December 18. Richland Creek December 24. King's Gap near Pulaski December 25. At Gravelly Springs, Ala., till March, 1865. Wilson's Raid on Macon, Ga., March 22-April 24. Near Elyton (Birmingham) March 28. Near Montevallo March 31. Ebenezer Church, near Maysville April 1. Selma April 2. Montgomery April 12. Columbus, Ga., April 16. Capture of Macon April 20. Duty in North Georgia and at Nashville, Tenn., till August. Mustered out August 11, 1865.

Regiment lost during service 7 Officers 58 Enlisted men killed and mortally wounded and 2 Officers and 179 Enlisted men by disease. Total 246.

10th Regiment, Missouri Cavalry

Overview:
Organized at Jefferson Barracks, Mo., October, 1862, from 28th Missouri Infantry. Bowen's Battalion assigned as Companies "A," "B," "C" and "D," and six Companies organized for 9th Missouri Cavalry assigned December 17, 1862, as Companies "E," "F," "G" and "H." Attached to District of St. Louis, Mo., to January,

1863. District of Memphis, Tenn., 16th Army Corps, Dept. of Tennessee, to March, 1863. Cavalry Brigade, District of Corinth, 16th Army Corps, to June, 1863. 3rd Brigade, 1st Cavalry Division, 16th Army Corps, to August, 1863. Cavalry Brigade, 15th Army Corps, to December, 1863. Winslow's Cavalry Brigade, 17th Army Corps, and District of Vicksburg to April, 1864. 2nd Brigade, 1st Cavalry Division, 16th Army Corps, to June, 1864. 2nd Brigade, Cavalry Division, Sturgis' Expedition, June, 1864. 2nd Brigade, 1st Cavalry Division, District of West Tennessee, to November, 1864. 1st Brigade, 4th Division, Cavalry Corps, Military Division Mississippi, to December, 1864. 2nd Brigade, Cavalry Division, District of West Tennessee, to February, 1865. 1st Brigade, 4th Division, Cavalry Corps, Military Division Mississippi, to May, 1865. 2nd Brigade, 4th Division, Cavalry Corps, Military Division Mississippi, to June, 1865.

Service:

Moved to Memphis, Tenn., December, 1862. Duty in the District of Memphis, Tenn., till February, 1863. Moved to Corinth, Miss., February 7-15. Actions at Glendale and Tuscumbia, Ala., February 22. Duty in that district till June. Courtney's Plantation April 11. Burnsville, Ala., and Glendale, Miss., April 14. Dodge's Expedition into Northern Alabama April 15-May 8. Barton Station April 16-17. Dickson Station, Great Bear Creek, Cherokee Station, and Lundy's Lane April 17. Dickson's Station April 19. Rock Cut near Tuscumbia April 22. Dickson's Station and Tuscumbia April 23. Leighton April 23. Lundy's Lane April 25, Town Creek April 27. Expedition from Burnesville to Tupelo, Miss., May 2-8. Guntown May 4. Tupelo May 5. Near Vicksburg, Miss., May 18 (Co. "C"). Expedition from Corinth to Florence, Ala., May 26-31. Florence, Ala., May 28. Hamburg Landing, Tenn., May 29-30. Iuka,

Miss., July 7. Jackson, Miss., July 29. Jacinto August 13. Expedition from Corinth to Henderson, Tenn., September 11-16. Clark's Creek Church September 13 (Detachment). Yazoo City, Miss., September 27. Expedition from Big Black River to Yazoo City September 27-October 1 (Detachment). Brownsville September 28. Canton September 28. Moore's Ford near Benton September 29. Messenger's Ford October 5. Expedition to Canton October 14-22. Brownsville October 15. Canton Road near Brownsville October 15-16. Treadwell's Plantation near Clinton and Vernon Cross Roads October 16. Bogue Chitto Creek October 17. Robinson's Mill near Livingston October 17. Livingston Road near Clinton October 18. Treadwell's Plantation October 20. Brownsville October 22. Near Yazoo City October 31. Operations about Natchez, Miss., December 1-10. Natchez December 10 (Detachment). Meridian Campaign February 3-March 2, 1864. Near Bolton's Depot and Champion's Hill February 4. Jackson February 5. Morton and Brandon February 7. Morton February 8. Meridian February 9-13. Hillsboro February 10. Meridian February 13-14. Lauderdale Springs February 16. Union February 21-22. Canton February 24. Near Canton February 26. Sharon February 27. Canton February 29. Livingston March 27. Near Mechanicsburg April 20. Ordered to Memphis, Tenn., April 29. Bolivar, Tenn., May 2. Sturgis' Expedition to Guntown, Miss., June 1-13. Rienzi, Miss., June 6. Danville, Miss., June 6. Brice's or Tishamingo Creek near Guntown June 10. Guntown June 24. Smith's Expedition to Tupelo, Miss., July 5-21. Tupelo July 14-15. Old Town Creek July 15. Smith's Expedition to Oxford, Miss., August 1-30. Tallahatchie River August 7-9. Hurricane Creek and Oxford August 9. Tallahatchie River August 10. Oxford August 12. Hurricane Creek August 13-14 and 19. Holly Springs August 27-28. Moved to Little Rock September 2-9.

Campaign against Price in Arkansas and Missouri September 17-November 30. Actions at Little Blue October 21. Big Blue and State Line October 22. Westport October 23. Engagement at the Marmiton or battle of Charlotte October 25. Osage Mine Creek, Marias des Cygnes, October 25. Rolla November 1. Expedition from Memphis to Moscow November 9-13. A detachment on Grierson's Raid on Mobile & Ohio Railroad December 21, 1864, to January 5, 1865. Verona December 25. Egypt Station December 28, 1864. Regiment at Louisville, Ky., till February, 1865. Moved to Gravelly Springs, Ala., February 5-15, 1865. Wilson's Raid from Chickasaw, Ala., to Macon, Ga., March 22-April 24. Near Montevallo, Ala., March 31. Ebenezer Church near Maplesville April 1. Selma April 2. Columbia, Ga., April 16. Capture of Macon, Ga., April 20. Duty at Macon and in Georgia till June. Mustered out June 20, 1865. (Co. "C" in demonstration on Haines' Bluff April 29-May 2, 1863. Siege of Vicksburg May 18-July 4. Advance on Jackson, Miss., July 5-10. Siege of Jackson July 10-17. Jackson July 29. Expedition to Yazoo City September 27-October 1.)

Regiment lost during service 2 Officers and 52 Enlisted men killed and mortally wounded and 3 Officers and 295 Enlisted men by disease. Total 352.

1st Regiment, Ohio Cavalry

Overview:

Organized at Camp Chase, Ohio, August 17-October 30, 1861. Left State for Louisville, Ky., December 9, 1861. Attached to 1st Division, Army Ohio, to October, 1862. (Cos. "F," "I," "K," "L" and "M" attached to 5th Division, Army Ohio, May to October, 1862.) Zahm's

2nd Brigade, Cavalry Division, Army Ohio, to November, 1862. (Cos. "F," "I," "K," "L" and "M" attached to 2nd Corps, Army Ohio, to November, 1862.) 2nd Brigade, Cavalry Division, Army of the Cumberland, to January, 1863. 2nd Brigade, 1st Cavalry Division, Army of the Cumberland, to March, 1863. 2nd Brigade, 2nd Cavalry Division, Army of the Cumberland, to October, 1864. 2nd Brigade, 2nd Division, Wilson's Cavalry Corps, Military Division Mississippi, to February, 1865. 2nd Brigade, 4th Division, Wilson's Cavalry, 1865. 1st Brigade, 4th Division, Wilson Corps Cavalry Corps, and Dept. of Georgia, to September, 1865.

Service:

Company "B" was at Headquarters of Gen. Mitchel in Kentucky October to December, 1861. Action at West Liberty, Ky., October 23. Rejoined Regiment at Louisville, Ky., December, 1861. Operations near Greensburg and Lebanon, Ky., January 28-February 2, 1862. Moved to Louisville, Ky., February 14, thence to Nashville, Tenn., February 28-March 3. Advance on Columbia March 14-15. Near Columbia March 15. March to Savannah, Tenn., March 28-April 7, thence moved to Pittsburg Landing, Tenn. Advance on and siege of Corinth, Miss., April 29-May 30. Pursuit to Booneville May 30-June 12. Reconnaissance toward Carrollville and Baldwyn June 3. Skirmish at Blackland June 3. Osborn's and Wolf Creeks, near Blackland, June 4 (Cos. "E," "I" and "M"). Guard duty along Memphis & Charleston Railroad till August. Near Russellville July 3 (Cos. "B" and "G"). Expedition to Decatur, Ala., July 12-16 (Detachment). Near Davis Gap July 12 (Detachment). Near Decatur July 15 (Co. "I"). Pond Springs July 24. Courtland and Trinity July 25 (Detachment). Moved to Dechard, Tenn., August 1. Salem August 6. Scout to Fayetteville August 17-20.

March to Louisville, Ky., in pursuit of Bragg August 21-September 25. Pursuit of Bragg into Kentucky October 1-22. Cedar Church, near Shepherdstown, October 3. Bardstown October 4. Battle of Perryville October 8 (Detachment). Pursuit of Bragg to Loudon October 10-22. Harrodsburg October 13. Stanford October 14. March to Nashville, Tenn., October 22-November 7. Duty there till December 26. Franklin December 12 and 26. Reconnaissance from Rural Hill December 20. Advance on Murfreesboro December 26-30. Nolinsville December 26. Near Murfreesboro December 29-30. Battle of Stone's River December 30-31, 1862, and January 1-3, 1863. Overall's Creek December 31, 1862. Shelbyville Pike January 5. Duty at Lavergne till June. Reconnaissance from Lavergne May 12. Middle Tennessee or Tullahoma Campaign June 23-July 7. Moore's Ford, Elk River, July 2. Occupation of Middle Tennessee till August 16. Expedition to Huntsville July 13-22. Passage of Cumberland Mountains and Tennessee River, and Chickamauga (Ga.) Campaign August 16-September 22. Reconnaissance from Stevenson to Trenton, Ga., August 28-31. Reconnaissance from Winston's Gap to Broomtown Valley September 5. Alpine, Ga., September 3 and 8. Reconnaissance from Alpine toward Lafayette, Ga., September 10. Alpine September 11. Battle of Chickamauga, Ga., September 19-21. Cotton Port Ford, Tennessee River, September 30. Operations against Wheeler and Roddy September 30-October 17. Greenville October 2. McMinnville October 4. Farmington October 7. Sim's Farm, near Shelbyville, October 7. At Paint Rock till November 18. Chattanooga-Ringgold Campaign November 23-27. Raid on East Tennessee & Georgia Railroad November 24-27. Charleston November 26. Cleveland November 27. March to relief of Knoxville, Tenn., November 28-December 8. Near Loudoun December 2. Expedition to

Murphey, N. C., December 6-11. Charleston and Calhoun December 28. Regiment re-enlisted January 4, 1864. Demonstration on Dalton, Ga., February 22-27, 1864 (Non-Veterans). Near Dalton February 23. Tunnel Hill, Buzzard's Roost Gap and Rocky Faced Ridge February 23-25. Tunnel Hill February 25. Buzzard's Roost February 27. Atlanta (Ga.) Campaign May 1-September 8, 1864. Decatur, Ala., May 26. Courtland Road, Ala., May 26. Pond Springs, near Courtland, May 27. Moulton May 28-29. Operations about Marietta and against Kennesaw Mountain June 10-July 2. McAffee's Cross Roads June 11. Noonday Creek June 15-19 and 27. Kenesaw Mountain June 21. Near Marietta June 23. Assault on Kennesaw June 27. Nickajack Creek July 2-5. Rottenwood Creek July 4. Chattahoochee River July 5-17. Raid to Covington July 22-24. Siege of Atlanta July 24-August 15. Garrard's Raid to South River July 27-31. Flat Rock Bridge and Lithonia July 28. Kilpatrick's Raid around Atlanta August 18-22. Flint River and Red Oak August 19. Jonesborough August 19. Lovejoy Station August 20. Operations at Chattahoochee River Bridge August 26-September 2. Occupation of Atlanta September 2. Operations against Hood and Forest in North Georgia and North Alabama September 29-November 3. Near Lost Mountain October 4-7. New Hope Church October 5. Dallas October 7. Rome October 10-11. Narrows November 11. Coosaville Road, near Rome, November 13. Near Summerville October 18. Little River October 20. Blue Pond and Leesburg October 21. Coosa River October 25. Ladiga, Terrapin Creek, October 28. Ordered to Louisville, Ky., and duty there till December. Ordered to Gravelly Springs, Ala., December 28, and duty there till March, 1865. Wilson's Raid to Macon, Ga., March 22-April 24. Near Montevallo March 31. Ebenezer Church April 1. Selma April 2. Montgomery April 12-13. Crawford and Girard April. Columbus and West Point

April 16. Capture of Macon April 20. Irwinsville, Ga., May 10. Capture of Jeff Davis. Duty in Georgia and South Carolina till September. Mustered out September 13, 1865.

Companies "A" and "C" ordered to West Virginia September 17, 1861. Attached to Army of Occupation, West Virginia, to October, 1861. Cheat Mountain District West Virginia, to January, 1862. Landers' Dlvision, Army Potomac, to March, 1862. Shields' 2nd Division, Banks' 5th Army Corps, and Dept. of the Shenandoah, to May, 1862. Cavalry, Shields' Division, Dept. of the Rappahannock, to June, 1862. Headquarters 2nd Corps, Army of Virginia, to September, 1862. Price's Cavalry Brigade, Military District of Washington, D. C., to March, 1863. 2nd Brigade, Stahel's Cavalry Division, 22nd Army Corps, Dept. of Washington, to June, 1863. Headquarters 3rd Division, Cavalry Corps, Army of the Potomac, to December, 1863. Defenses of Washington, D. C., to January, 1864.

Participating in skirmish at Bloomery Gap, Va., February 4, 1862. Advance on Winchester March 7-15. Battle of Winchester March 23. Occupation of Mt. Jackson April 17. Battle of Cedar Mountain August 9. Pope's Campaign in Northern Virginia August 16-September 2. Catlett's Station August 22. Centreville August 27-28. Groveton August 29. Bull Run August 30. Chantilly September 1. Duty in Defenses of Washington till June, 1863. Battle of Gettysburg, Pa., July 1-3, 1863. Monterey Gap July 4. Emmettsburg July 5. Hagerstown July 6-12. Falling Waters July 14. Hartwood Church August 28. Advance from the Rappahannock to the Rapidan September 13-17. Bristoe Campaign October 9-22. Hartwood Church November 5. Mine Run Campaign November 26-December 2. In

Defenses of Waghington, D. C., till January, 1864, when rejoined Regiment.

Regiment lost during service 6 Officers and 45 Enlisted men killed and mortally wounded and 3 Officers and 150 Enlisted men by disease. Total 204.

7th Regiment, Ohio Cavalry

Overview:

Regiment organized at Ripley, Ohio, October, 1862, and duty there till December, 1862. First Battalion (Cos. "A," "B," "C," "D"), ordered to Lexington, Ky., November 22, 1862, and duty there till December 21. 2nd Battalion ordered to Lexington, Ky., December 20, 1862, and 3rd Battalion to same point December 31, 1862. Attached to District of Central Kentucky, Dept. Ohio, to January, 1863. 2nd Brigade, District of Central Kentucky, to April, 1863. 1st Provisional Cavalry Brigade, 23rd Army Corps, Army Ohio, to June, 1863. 3rd Brigade, 1st Division, 23rd Army Corps, to August, 1863. 3rd Brigade, 4th Division, 23rd Army Corps, to November, 1863. 1st Brigade, 2nd Division, Cavalry Corps, Dept. Ohio, to May, 1864. 1st Brigade, Cavalry Division, District of Kentucky, 5th Division, 23rd Army Corps, to July, 1864. 1st Brigade, Cavalry Division, 23rd Army Corps, to August, 1864. Mounted Brigade, Cavalry Division, 23rd Army Corps, to September, 1864. 2nd Brigade, Cavalry Division, 23rd Army Corps, to November, 1864. 2nd Brigade, 6th Division, Wilson's Cavalry Corps, Military Division Mississippi, to December, 1864. 1st Brigade, 6th Division, Cavalry Corps, to February, 1865. 2nd Brigade, 4th Division, Cavalry Corps, to July, 1865.

Service:

1st Battalion participated in Carter's Raid into East Tennessee and Southwest Virginia December 21, 1862, to January 5, 1863. Passage of Moccasin Gap December 29, 1862. Actions at Zollicoffer Station December 30. Watauga Bridge, Carter's Station, December 30 (Cos. "A," "D"). Regiment participated in operations in Central Kentucky against Cluke's forces February 18-March 5, 1863. Slate Creek near Mt. Sterling February 24 and March 2. Operations against Pegram March 22-April 1. Dutton's Hill March 30. Expedition to Monticello and operations in Southeast Kentucky April 26-May 12. Monticello May 1. Rocky Gap, Monticello, June 9. Carter's Raid in East Tennessee June 16-24. Knoxville June 19-20. Roger's Gap June 20. Powder Springs Gap June 21. Scout to Creelsborough June 28-30. Pursuit of Morgan July 1-25. Buffington Island, Ohio, July 19. Operations against Scott in Eastern Kentucky July 26-August 6. Near Rogersville July 27. Richmond July 28. Lancaster July 31. Paint Lick Bridge July 31. Lancaster August 1. Burnside's Campaign in East Tennessee August 16-October 19. Winter's Gap August 31. Expedition to Cumberland Gap September 4-9. Capture of Cumberland Gap September 9. Carter's Station September 22. Zollicoffer September 24. Jonesboro September 28. Blue Springs October 5 and 10. Sweetwater October 10-11. Pursuit to Bristol October 11-17. Blountsville October 13-14. Moved to Rogersville October 17-19. Knoxville Campaign November 4-December 23. Action at Rogersville November 6. Stock Creek November 14. Defense of Cumberland Gap during siege of Knoxville November 17-December 5. Morristown December 10. Cheek's Cross Roads December 12. Russellville December 12-13. Bean's Station December 14. Rutledge December 16. Blain's Cross Roads December 16-19. Rutledge December 18. Stone's Mill December 19. New Market December 23. Dandridge December 24. Mossy Creek

December 26. Operations about Dandridge January 16-17, 1864. Kimbrough's Cross Roads January 16. Dandridge January 17. Operations about Dandridge January 26-28. Fair Garden January 27. Ordered to Nicholasville, Ky., February. Operations against Morgan in Kentucky May 31-June 20 (Detachment). Cynthiana June 12. March to Atlanta, Ga., July 4-26. Siege of Atlanta July 26-September 2. Stoneman's Raid to Macon July 27-August 6 (Co. "D"). Clinton and Macon July 30 (Co. "D"). Hillsborough July 30-31 (Co. "D"). Sandtown and Fairburn August 15. At Decatur till October 4. At Atlanta till November 6. Moved to Nashville, Tenn. Nashville Campaign November-December. Henryville November 23. Columbia, Duck River, November 24-27. Duck River Crossing November 28. Columbia Ford November 29. Franklin November 30. Nashville December 15-16. Pursuit of Hood to the Tennessee River December 17-28. West Harpeth River and Franklin December 17. Spring Hill December 18. Richland Creek December 24. Pulaski December 25-26. Moved to Gravelly Springs, Ala., and duty there till March, 1865. Wilson's Raid from Chickasaw, Ala., to Macon, Ga., March 22-April 24. Montevallo March 31. Ebenezer Creek near Mapleville April 1. Selma April 2. Montgomery April 12. Columbia, Ga., April 16. Capture of Macon April 20. Scout duty in Northern Georgia till May 15. Moved to Nashville, Tenn., and duty there till July. Mustered out July 4, 1865.

Regiment lost during service 2 Officers and 26 Enlisted men killed and mortally wounded and 4 Officers and 197 Enlisted men by disease. Total 229.

CONFEDERATE ALABAMA TROOPS

4th Regiment, Alabama Cavalry

Overview:

    4th (Roddey's) Cavalry Regiment was organized at Tuscumbia, Alabama, in October, 1862, and moved to Tennessee where it wintered. The men were from Franklin, Lauderdale, Lawrence, and Walker counties. During the next spring it was sent to Northern Alabama, assigned to General Roddey's Brigade, then took an active part in raiding and attacking the Federals. In April, 1864, the regiment was transferred to the Department of Alabama, Mississippi, and East Louisiana. After fighting at Brice's Cross Roads it saw action in various conflicts from Montevallo to Selma where on April 2, 1865, most of the unit was captured. The remaining part surrendered at Pond Spring. Its commanders were Colonels William A. Johnson and Philip D. Roddey, Lieutenant Colonel E.M. Windes, and Majors R.W. Johnson and John E. Newsom.

6th Regiment, Alabama Cavalry

Overview:

    6th Cavalry Regiment was organized at Pine Level, Alabama, during the spring of 1863. It contained men from Montgomery, Coffee, Tallapoosa, Pike, Barbour, Macon, Henry, and Coosa counties. The unit was brigaded under General Clanton, served for a time in Florida, then in August became part of the garrison at Montgomery. Later it was attached to Armstrong's command and saw action in various conflicts during the Atlanta Campaign. In August, 1864, the regiment returned to Clanton's Brigade and fought at Bluff Springs, Florida, and in southern Alabama. With less than 200 men it was included in the surrender of the Department of Alabama, Mississippi, and East Louisiana. Its commanders were Colonel C.H. Colvin, Lieutenant Colonel Washington T. Lary, and Major E.A. McWhorter.

7th Regiment, Alabama Cavalry

Overview:

7th Cavalry Regiment was formed at Newborn, Alabama, during July, 1863, with companies were raised in the counties of Randolph, Shelby, Greene, Pickens, and Montgomery. For a year, the unit served in the Pollard area assigned to General Clanton's Brigade. In July, 1864, it contained 451 men, but was not serving as one command; two companies were with General Page, and eight rode with Colonel I.W. Patton. The 7th was later attached to B.M. Thomas', W.W. Allen's, and Bell's Brigade. It took part in the raid on Johnsonville and was engaged in the fighting as Hood moved toward Nashville. In April, 1865, it had less than 300 effectives and half that number surrendered at Gainesville, Alabama, in May. The field officers were Colonel Joseph Hodgson, and Lieutenant Colonels Turner Clanton, Jr., Henry J. Livingston, and F.C. Randolph.

8th Regiment, Alabama Cavalry (Hatch's)

Overview:

8th (Hatch's) Cavalry Regiment was organized at Newbern, Alabama, in April, 1864, by adding one company to the nine of Hatch's Battalion that had entered Confederate service the previous winter. The men were form Sumter, Dallas, Tuscaloosa, Greene, Marengo, Choctaw, and Fayette counties. It joined General Pillow at Blue Mountain, then was assigned to C.G. Armistead's Brigade, Department of Alabama, Mississippi, and East Louisiana. The unit fought at Lafayette, Georgia, and sustained 105 casualties. Later it moved to Selma, served at Talladega, and in September totaled 241 effectives. It went on to confront the Federals in Florida and Alabama, and on May 14, 1865, surrendered at Gainesville. Its commanders were Colonels Charles P. Ball and Lemuel D. Hatch, and Majors William T. Poe and Richard H. Redwood.

8th Regiment, Alabama Cavalry (Livingston's)

Overview:

8th (Livingston's) Cavalry Regiment completed its organization at Gadsden Alabama, during the spring of 1864. In July it reported to General Pillow at Blue Mountain with about 250 men. Later the unit was assigned to General Clanton's Brigade, Department of Alabama, Mississippi, and East Louisiana. It was engaged at Ten Islands, served in Western Florida, then sustained heavy losses in the conflict at Bluff Springs. By February, 1865, its strength had grown to 600, but after skirmishing in Alabama, few were present when the regiment surrendered at Gainesville in May. Colonel Henry J. Livingston, Lieutenant Colonel T.L. Faulkner, and Major Sidney A. Moses were in command.

CONFEDERATE TENNESSEE TROOPS

3rd Regiment, Tennessee Cavalry

Overview:
>
> 3rd (Forrest's Old) Cavalry Regiment was organized at Memphis, Tennessee, in October, 1861, as an eight-company battalion. In January, 1862, it was increased to regimental size. The history of this unit is very complex because over twenty companies from Tennessee, Kentucky, Texas, Mississippi, and Louisiana were attached to it at one time or another. After the Battle of Shiloh, four companies were transferred to the 4th (Russell's) Alabama Cavalry Regiment and thereafter it was called 18th or 26th Battalion, Balch's Battalion, and McDonald's Battalion. It fought at Fort Donelson and Shiloh, and during April, 1862, contained 463 effectives. The unit was attached to Forrest's, F.C. Armstrong's, and E.W. Rucker's Brigade, and served in the Army of Tennessee and the Department of Alabama, Mississippi, and East Louisiana. It confronted the Federals in Tennessee, Mississippi, Kentucky, and Alabama, and surrendered in May, 1865. The field officers were Colonels Nathan B. Forrest and David C. Kelley; Lieutenant Colonels P.T. Allin, Robert M. Balch, and Edward E. Porter; and Majors James C. Blanton, William H. Forrest, Charles McDonald, and Edwin A. Spotswood.

20th Regiment, Tennessee Cavalry (Russell's)

Overview:
>
> 20th (Russell's) Cavalry Regiment [also called 15th Regiment] was organized in February, 1864. Its members were recruited in the counties of Henry, Gibson, Carroll, Madison, Dyer, Humphreys, and Weakley. The unit was placed in T.H. Bell's Brigade, Department of Alabama, Mississippi, and East Louisiana, and fought at Okolona, Brice's Cross Roads,

and Harrisburg. Later it skirmished in Tennessee, was part of Hood's operations, then moved to Mississippi. The regiment ended the war in Alabama and on May 3, 1865, contained 29 officers and 217 men. The field officers were Colonel Robert M. Russell, Lieutenant Colonel Henry C. Greer, and Major H.F. Bowman.

CONFEDERATE TEXAS TROOPS

10th Regiment, Texas Cavalry (Locke's)

Overview:

10th Cavalry Regiment was organized with about 900 men during the late summer of 1861. Many of its members were recruited in the towns of Quitman and Tyler, and the counties of Upshur, Rusk, and Cherokee. For the first few months it served in Texas, Arkansas, and Louisiana, then was dismounted after crossing the Mississippi River. After fighting at Richmond, the unit was assigned to General Ector's Brigade in the Army of Tennessee. It participated in numerous battles from Murfreesboro to Atlanta, endured Hood's winter operations in Tennessee, and aided in the defense of Mobile. This regiment totaled 565 effectives during the spring of 1862 and lost thirty-four percent of the 350 engaged at Murfreesboro. Very few surrendered on May 4, 1865. The field officers were Colonels James M. Barton and W.D. Craig, and Majors Wiley B. Ector and Hulum D. E. Redwine.

BIBLIOGRAPHY

GOVERNMENT DOCUMENTS:

Alabama Civil War Archives. Confederate Inspection Reports, Clanton's Battery Alabama Artillery Inspection Reports, LPR-79, Container 9.

_____. 3d and 4th Reserve Regiments, LPR-79, SG-25070.

_____. Home Guards, LPR-79, Container 8.

_____. Waddell's 20th Artillery Battalion, SPR-398.

_____. Bellamy's Artillery, SG-25094.

_____. Emory's Artillery, SG-25095.

_____. 7th Alabama Cavalry, RH-27, SG-24912.

_____. Russell County Units, RH-46, SG-24931.

_____. Maps of Wilson's Raid, B-349, B-353 & CB-95.

Columbus Hospital Reports (April 16-22, 1865). Post Hospital Log. Mayor's Office, Columbus, Georgia Government Center, Columbus, Georgia.

Columbus Sexton's Reports, Book E, 1853-1866, City Council records, City Clerk's Office. Columbus, Georgia Government Center, Columbus, Georgia.

Compiled Service Records of Confederate Soldiers Who Served in Organizations from the State of Alabama. M311, Index Roll M374. Roll 70, Clanton's Artillery Battery, Rolls 23-24, 7th Cavalry, Rolls 68-69, 20th Battalion, Light Artillery.

Compiled Service Records of Confederate Soldiers Who Served in Organizations from the State of Georgia. National Archives, M266: 3d Cavalry, rolls 16-19; 1st. Georgia Regulars (Ramsey's), Rolls 143-45; 2nd Georgia Infantry, Rolls 152-56; 15th Georgia Infantry, Rolls 290-96; 17th Georgia Infantry, Rolls 303-07; 20th Georgia Infantry, Rolls 325-32; Index, M266, Georgia, Rolls 1-67.

Compiled Records Showing Service of Military Units in Confederate Organizations. National Archives, M861: Rolls14 and 16.

Confederate General Staff Officers and non-Regimental Enlisted Men. National Archives, M331, Roll 22.

Confederate List of Volunteer Companies (Georgia). Georgia Department of Archives and History, Miscellaneous, Drawer 283, Roll 58.

Confederate States of America Collection. War Department, List of Official Communications, box 110, Maps, Library of Congress, Washington, D.C.

Governor Joseph E. Brown's Incoming Correspondence, 1861-1865. Georgia Department of Archives and History, Atlanta, Georgia.

Muscogee County Civil War Pension Records, Widow of John Lindsay. Columbus, Georgia Government Center Archives.

_____. *Court Records* (Local Troop Rosters- Probate Court), Columbus, Georgia, Government Center Archives.

_____. *Statistics* (Muscogee County, Georgia- Mayor's Office), Columbus, Georgia, Government Center Archives.

National Archives, Civil War Claims, Record Group 217, Microfiche # M1658, Fiche # 1, 2, & 3; Claim Number 15615.

U.S. War Department, *The War of the Rebellion: A Compilation of the Official Records of the Union and Confederate Armies*. 127 Volumes, index and atlas. Washington D.C., 1880-1901.

Wright, Marcus J., *List of Field Officers, Regiments, and Battalions in the Confederate States Army.* J.W. Burke Company, Macon, Georgia, 1912.

ARCHIVAL DOCUMENTS:

Columbus State University. SMC-16, *Post Register of Sick & Wounded in Hospitals*, Columbus, Georgia, 1864-1865. CSU Archives, Simon Schwob Library, Chattahoochee Valley Historical Collection, Columbus, Georgia

Comer, Mrs. James. Diary, 1846-1865. Unpublished manuscript. Columbus State University Archives.

Conzett, Josiah. *My Civil War, Before, During and After.* Unpublished manuscript. Aurhor's Collection.

Crowell, Andrea, Norman Bates, Data and photograph. Greene Collection.

Garrard, Leonard, Personal archives, genealogical notes, family papers, and interview with author. Greene Collection.

Harrington, Benjamin F. The Civil War Diaries of Benjamin F. Harrington: 4th Iowa Cavalry, 2nd Regiment, Company E. Unpublished Diary: Greene Collection.

Hoole, William Stanley. Ed., Seventh Alabama Cavalry Regiment. Unpublished transcript, author unknown, Alabama Archives.

Jackson, Alphonza. Diary and Letters. Box 76-7, Microfilm Library, Georgia Department of Archives and History, quoted in Bragg, p. 107.

Jones, Louise Gunby. Some Reminiscences and Incidents of the Last Battle of the Civil War at Columbus, Easter, April 16, 1865. Columbus State University Archives.

Moore, Lindsey Rev. B., N.H. Clanton data and photograph. Greene Collection.

Saltmarsh, T. William, Jr. Letter dated July 2, 2002 (descendant of Theodore Moreno), Greene Collection.

Smith, Jenny, MC-9, Box 2, Folder 3, Columbus Museum of Arts & Sciences Collection, CSU Archives, Simon Schwob Library, Chattahoochee Valley Historical Collection, Columbus, Georgia.

Stanford, Brant, Letter dated July 24, 2002. William Stanford data. Greene Collection.

Swift, Charles. The Last Battle of the Civil War, Gilbert Printing Co., Columbus, Georgia, 1915.

U.S. Army Military Institute, Carlisle, Pennsylvania. Photos and archival data. Greene Collection.

Waddell, James D., Papers, Special Collections, Robert W. Woodruff Library, Emory University, Atlanta, Georgia.

NEWSPAPERS

Columbus (Georgia*) Daily Enquirer*
Columbus (Georgia) *Daily Sun*
Columbus (Georgia) *Weekly Enquirer*
Columbus (Georgia) *Times and Sentinel*
Russell Register (Alabama)

PUBLISHED CAMPAIGN AND BATTLE NARRATIVES

Dinkins, CPT James. "The Last Campaign of Forrest's Cavalry." *Confederate Veteran.* Vol. XXXV, April 1927, p. 136-138.

Grant, W. W. Dr. "Recollections of the Last Battle." *Confederate Veteran.* April 1915, Vol. 23, p. 163-165.

Gilpin, E. N., "The Last Campaign- A Cavalryman's Journey," *Journal of the United States Cavalry Association*, April 1908, p. 617-675, Ketcheson Press, Leavenworth, Kansas.

Johnson, Robert Underwood and Clarence Buel. eds. *Battles and Leaders of the Civil War*. 4 vols., New York: The Century Co., 1884-1888.

Yeoman, J.A.O. Captain. "The Wilson Raid Through Northern Alabama and Georgia." p. 222-23. Curry, W.L., *Four Years in the Saddle: History of the First Regiment Ohio Volunteer Cavalry. War of the Rebellion 1861-1865.* Freedom Hill Press: Jonesboro, Georgia, 1984.

PUBLISHED SOURCES

Aimore, Alan Conrad. *The Official Records of the American Civil War; A Researcher's Guide.* West Point, NY: United States Military Academy, 1972.

Albaugh, William A., *Confederate Edged Weapons.* New York: Harper & Brothers, 1960.

Avery, Isaac W., *The History of the State of Georgia, From 1850 to 1881.* New York: Brown & Derby Publishers, 1881.

Barfield, Louise Calhoun. *History of Harris County, Georgia.* Columbus, Georgia: Columbus Office Supply, 1961.

Bergeron, Arthur W. Jr. *Guide to Louisiana Confederate Military Units, 1861-1865.* Baton Rouge: Louisiana State University Press, 1988.

Bragg, William Harris. *Joe Brown's Army: The Georgia State Line, 1862-1865* Macon, Georgia: Mercer Press, 1987.

Candler, Allen D. comp. *The Confederate Records of the State of Georgia.* 6 vols. (vol. 5 never published). Atlanta: State Printing Office, 1909-11.

Coulter, Harold S. *A People Courageous: A History of Phenix City, Alabama.* Columbus, Georgia: Howard Printing Company, 1976.

Crumpton, Washington Bryan. *A Book of Memories 1842-1920.* Montgomery, Alabama: Baptist Mission Board, 1921.

Crute, Joseph H. Jr., *Units of the Confederate States Army.* Midlothian, Virginia: Derwent Books, 1987.

Curry, W.L. *Four Years in the Saddle: History of the First Regiment Ohio Volunteer Cavalry. War of the Rebellion 1861-1865.* Jonesboro, Georgia: Freedom Hill Press, 1984.

Dameron, J. David. *Benning's Brigade, Vol. 1, A History and Roster of the Fifteenth Georgia.* Spartanburg, South Carolina: Reprint Company, 1997.

_____. *General Henry Lewis Benning: This Was a Man.* Heritage Books: Bowie, Maryland, 2003.

Dyer, Frederick H., *A Compendium of the War of the Rebellion, Vol. III,* New York: Yoseloff, 1959.

Ellingson, Paul. ed. *Confederate Flags in the State Capitol Collection.* Atlanta: Georgia Office of Secretary of State, 1994.

Evans, General Clement A. ed. *Confederate Military History; a library of Confederate States history, written by distinguished men of the south.* Vol. 6, Georgia. Atlanta, Ga.: Confederate Publishing Company, 1899.

Fielder, Herbert. *A Sketch of the Life and Times and Speeches of Joseph E. Brown.* Springfield, Mass.: Springfield Printing Co., 1883.

Fox, William F. *Regimental Losses in the American Civil War,* 1861-1865. Albany: Albany Publishing Co., 1889.

Georgia Division. United Daughters of the Confederacy. *Confederate Reminiscences and Letters 1861-1865, Volume I-XV.* Atlanta, Georgia: United Daughters of the Confederacy, 1998.

Henderson, Lillian. comp. *Roster of the Confederate Soldiers of Georgia*, 1861-1865. 6 vols. Hapeville, Georgia: Longina & Porter, 1959-1964.

Houghton, W. R., and M. B. Houghton, *Two Boys in the Civil War and After.* Montgomery, Alabama: The Paragon Press, 1912.

Jones, James Pickett. *Yankee Blitzkrieg: Wilson's Raid through Alabama and Georgia*. Athens: The University of Georgia Press, 1976.

Jordan, General Thomas and J. P. Jordan. *The Campaigns of Lieutenant General N. B. Forrest and of Forrest Cavalry.* New Orleans: Blelock & Company, 1868.

Keenan, Jerry. *Wilson's Cavalry Corps: Union Campaigns in the Western Theatre, October 1864 through Spring 1865.* Jefferson, North Carolina: McFarland & Company, 1998.

Larson, James. *Sergeant Larson, 4th Cavalry*. San Antonio: Southern Literary Institute, 1935.

Mandrell, Regina Moreno Kirchoff. *Our Family: Facts and Fancies. The Moreno and Related Families*, Pensacola, Florida: Perdido Bay Press, 1988.

Martin, *Columbus, Georgia, From Its Selection as a "Trading Town" in 1827, To Its Partial Destruction by Wilson's Raid in 1865.* Columbus, Georgia: Thomas Gilbert Publishing, 1874.

Mears and Company. comp. *The Columbus Directory for 1859-1860*. Columbus, Georgia: Sun Book and Job Printing Office, 1859.

Michie, Peter S., *The Life and Letters of Emory Upton, Colonel of the Fourth Regiment of Artillery, and Major General, U.S. Army.* New York: D. Appleton and Company, 1885.

Mills, Charles K. *Harvest of Barren Regrets: The Army Career of Frederick William Benteen 1834-1898.* Glendale, California: Arthur Clark Company, 1985.

Mosocco, Ronald A. *The Chronological Tracking of the American Civil War Per the Official Records of the War of the Rebellion.* Williamsburg, Va.: James River Publications, 1994.

Nisbet, John W. *Four Years on the Firing Line.* Wilmington, North Carolina: Broadfoot Publishing Co., 1987.

Northern, William J. *Men of Mark in Georgia. A Complete and Elaborate History of the State . . . Chiefly Told in Biographies . . . 7 vols.*, Atlanta: A.B. Caldwell, 1907-1912.

Phillips, Ullrich Bonnell. *Toombs, Stephens, Cobb Correspondence., American History Association Annual Report, 1913.* part II, for the Year 1911. Washington D. C.: Government Printing Office, 1913.

_____. *The Life of Robert Toombs.* New York: The Macmillan Company, 1913.

Pickett, Albert James. *History of Alabama and Incidentally of Georgia and Mississippi, from the Earliest Period.* Tuscaloosa, Alabama: Willow Publishing Company, 1962.

Rogers, William Warren. *Confederate Home Front: Montgomery During the Civil War.* Tuscaloosa: University of Alabama Press, 1999.

Scott, William Forse. *The Story of a Cavalry Regiment: The Career of the Fourth Iowa Veteran Volunteers, From Kansas to Georgia, 1861-1865.* New York: G.P. Putnam's Sons, 1893.

Sifakis, Stewart. *Compendium of the Confederate Armies: Alabama.* New York: Facts on File, 1992.

_____. *Compendium of the Confederate Armies: Louisiana.* New York: Facts on File, 1992.

_____. *Compendium of the Confederate Armies: Missouri.* New York: Facts on File, 1992.

_____. *Compendium of the Confederate Armies: South Carolina and Georgia.* New York: Facts on File, 1995.

Smith, John T., "Grant Family has deep roots in Russell County. *The Phenix Citizen Good News*, Februrary 28, 2002.

Standard, Diffee William. *Columbus, Georgia in the Confederacy: The Social and Industrial Life of the Chattahoochee River Port.* New York: William Frederick Press, 1954.

Taylor, Richard. *Destruction and Reconstruction: Personal Experiences of the Late War.* New York: D. Appleton and Company, 1879.

Telfair, Nancy. *History of Columbus, Georgia, 1828-1928.* Columbus, Georgia: Historical Publishing Company, 1929.

Turner, Maxine T. *Navy Gray: A Story of the Confederate Navy on the Chattahoochee and Apalachicola Rivers.* Tuscaloosa, Alabama: The University of Alabama Press, 1988.

Uzar, Sandra White. "Confederate Hospitals in Columbus, Georgia," *Muscogiana,* Volume 3, Numbers 1 & 2, (Spring 1992), pp. 16-21.

Walker, Anne Kendrick. *Russell County in Retrospect: An Epic of the Far Southeast,* Richmond, Virginia: Dietz Press, 1950.

Wells, Tom Henderson. *The Slave Ship Wanderer.* Athens: The University of Georgia, 1967.

Wills, Brian Steel. *A Battle from the Start.* New York: Harper Collins, 1992.

Wilson, James Harrison. *Under the Old Flag,* New York: D. Appleton and Company, 1912.

Worsley, Etta Blanchard. *Columbus on the Chattahoochee.* Columbus, Georgia: Columbus Office Supply Company, 1951.

Index

A

Alabama

_____ Bluff Springs 32, 118, 257

_____ Decatur 12, 101, 264

_____ Elyton (Birmingham) 34, 46, 265

_____ Girard 187-188,191-194,196,200,209,229

_____ Gravelly Springs 21

_____ Montevallo 34-35, 46, 103, 256

_____ Montgomery 2, 5-6. 12, 16, 21-22, 43-51, 57, 96, 101-102, 109, 230

_____ Phenix City iv, v, 3,16,55,57,165,231,290

_____ River 21,41,43-45

_____ Selma 2,4,12,16, 31-33, 35, 41, 43, 46-47, 52, 74, 101, 119, 134, 158, 163, 177, 195

Alabama Troops (pp. 201-207)

_____ 2nd Alabama Cavalry Regiment

_____ 3rd Alabama Cavalry Regiment

4th Alabama Cavalry Regiment 275

6th Alabama Cavalry Regiment 115,119,208,275

7th Alabama Cavalry Regiment 119-120,122,174,204,227,239,276, 283

8th Alabama Cavalry Regiment 119,203,278

_____ 3rd Alabama Reserves

_____ 4th Alabama Reserves

_____ 5th Alabama Reserves

_____ 15th Alabama Infantry

_____ 18th Alabama Infantry

_____ 20th Alabama Infantry

_____ 27th Alabama Infantry

_____ 57th Alabama Infantry

_____ Cole's Alabama Battalion

_____ Clanton's Alabama Artillery Battery

_____ Waddell's Alabama Artillery Battery

Adams, Daniel 28,37,45-47,110,115,116,119,166,171,181

Alexander, Andrew 39, 63, 131, 137, 145, 171, 250

Arkansas, 9th Infantry 201-207

B

Bates, Norman 170, 227

Benteen, Colonel Frederick 143, 251, 293

Brown, Joseph (Governor) 23,82,91-92, 109, 112

Buford, Abraham 2,4,6, 26,29-34, 37, 44-47, 51-52, 102-103, 110, 119-120, 130, 165-166, 171, 174, 181, 197, 232, 239, 249

C

Canby, Edward 32,43,47,101

Chalmers, James 25-26, 37

Chattahoochee River 2, 5, 8-9, 16, 180, 191-193, 217, 230-232, 271, 286, 297

Chattahoochee Valley iv, 13, 181, 196, 237

Chehaw Station 12

Clanton, James 32-33, 73, 119, 250

Clanton, Nathan 117, 124,149,150, 154, 162-163, 165, 175, 180, 203, 251

Cobb, Howell 1,4,6,23,28,46,63,80-82, 92, 98-99, 101-102, 108, 111, 166, 171, 174, 187-188

Columbus, Georgia iv-v,1-9,12-13,16,31,46-49,52-124,130-140,174-199

Confederate Casualties 201-209

Confederate Troops

4th Alabama Cavalry Regiment 275

6th Alabama Cavalry Regiment 115,119,208,275

7th Alabama Cavalry Regiment 119-120,122,174,204,227,239,276, 283

8th Alabama Cavalry Regiment 119,203,278

Crossland, Edward 26, 37

Croxton, John T. 34,36

CSS Chattahoochee 77

CSS Jackson 77-78, 114, 129, 180, 191, 223, 244

CSS Muscogee 77

CSS Shamrock 77

CSS Viper 77

Crumpton, Washington 114,129,166-169,220

D

Duck River 265,275

E

Ebenezer Church 1,35-36,40,103,135,152,230

F

Flag (Confederate) 127

_____ Union 126

Forrest, Nathan Bedford 1,11,15,24-48,65-66,101,103,109,119,187-183,230

Fort, fortification 7,31,41-42,46,50-51,63,85-88,96,101,106,110,112,114-127,131,137, 142-145,152-154,163-167

G

Gammill, John 36

Georgia, Atlanta 12-13,31,74,94-95,116,177,257

Georgia, Columbus iv-v,1-9,12-13,16,31,46-49,52-124,130-140,174-199

Georgia, Macon iv,3-4,57,74,101-102,109,116-117,170,181-182,188,196-198

Georgia State Line (1st and 2nd) 91-94,109-110,118,123,159-161,171,197,202-203

Georgia Troops (pp. 201-207)

_____ 2nd Georgia Cavalry

_____ 3rd Georgia Reserves

_____ 9th Georgia Artillery Battalion

_____ 12th Georgia Cavalry

_____ 25th Georgia Reserves

_____ 46th Georgia Infantry

_____ Georgia Arsenal Battalion

_____ Pemberton's Georgia Cavalry Company

Grant, James 165,180,204

Grant, U.S. 12-16,21,43,46

Guillet, Isidore 95-96,165,209

H

Haiman Brothers 72-74

_____ Sword 190

_____ Pistol 190

Henricks, Charles 50-52,142,148,152,154-156,171,182-184,194-196,216

Hood, John Bell 21

Hosea, Lewis 26-27

I

Iowa Troops

3rd Iowa Cavalry Regiment 177,227,239

4th Iowa Cavalry Regiment 44,50,52,140,143,146,148,154,157-160,166-168,171,174,177, 184,212,227,258

5th Iowa Cavalry Regiment 262

J

Jackson, Alphonza 160

Jackson, William 26-27, 35-36, 249

Jackson's Station 109

Jacques Battalion 89-90, 122, 202-203, 208

K

King Philip 66

Kinsel, Charles 90

L

Lee, Robert E. 229

Long, Eli 31, 37-38, 50, 53, 250

M

McCook, Edward 31, 36, 46, 49, 187, 250, 264

Mississippi 13-15, 21, 26, 28, 41, 44, 47, 71, 116-118, 204, 206, 229, 254

Mississippi Troops (pp. 201-207)

_____14th Mississippi Infantry Regiment

_____23rd Mississippi Infantry Regiment

_____31st Mississippi Infantry Regiment

_____44th Mississippi Infantry Regiment

Missouri Troops 51,239,241,243,265

10th Missouri Cavalry Regiment 50-51,239,241,243,265

N

Navy Yard (Confederate) 113

O

Ohio Troops 52,133,174,241,268,273, 291

1st Ohio Cavalry Regiment 133,174,241,268,291

7th Ohio Cavalry Regiment 52,174,273

Osnaburg 61,189

P

Pemberton, John 91,120,124,165,204

Q

R

Race Pass 180

Red Hill 106,115-117,124,137,145,152,154-156,161-165

Roddey, Philip 26-32, 34-37, 251

Rousseau, Lovell 12-13

S

Sandfort Road 137

Steele, Frederick 32

Summerville Road 8,115,117,123,129,139-140,142-145,147,153-155,157-158

T

Taylor, Richard 21-23,27,33,38,40-41,46,101,103,115,181,294

Tennessee 1,12,15,21,24-25,28,30,70-71,95,99,101,107,116,118,162,186,203

_____ Franklin 15,18,261,265

_____ Nashville 12,15,18,75,263,265

Tennessee Troops (pp. 201-207)

_____ 29th Tennessee Infantry

_____ 40th Tennessee Infantry

_____ Bates Escort Company (Cavalry)

Texas Troops (pp. 201-207)

_____ 10th Texas Cavalry

_____ 24th Texas Infantry

U

Union Casualties 210-214

Union Troops

3rd Iowa Cavalry Regiment 177,227,239

4th Iowa Cavalry Regiment 44,50,52,140,143,146,148,154,157-160,166-168,171,174,177, 184,212,227,258

5th Iowa Cavalry Regiment 262

1st Ohio Cavalry Regiment 133,174,241,268,291

7th Ohio Cavalry Regiment 52,174,273

10th Missouri Cavalry Regiment 50-51,239,241,243,265

Upton, Emory 3,31,34-36,39-42,49-52,100,131,133-135,138-145,147-148,152,163,174, 178,187,220

V

Von Zinken, Leon 94-95,97-98,102-105,108-112,114-116,118,121,129-131,162,165,174,181,209,232,234

W

Waddell, James 115-117,124,135,137-138,154,164,175,202-207

Watts, Thomas (Governor) 28,32,46,93, 108,110

Wilson, James H. 1,13,15-20,33,39-41,47,51,100-103,113,119,131,139-142,148-150,161-163,174,187,199

X

Y

Yankees 66,97,104-105,110,131,159,169,184,193,196-198

Z

www.ingramcontent.com/pod-product-compliance
Lightning Source LLC
Chambersburg PA
CBHW071856290426
44110CB00013B/1172